P9-APU-113

A SCIENTIFIC AMERICAN *Book*

EVOLUTION

W. H. FREEMAN AND COMPANY
San Francisco

The Cover

The photograph on the cover symbolizes the theme of this issue of SCIENTIFIC AMERICAN: evolution. It shows two fossilized fish of the Eocene epoch; they are members of the herring family, a group of fishes that probably evolved during late Mesozoic times. The first of the vertebrates to evolve, early in Paleozoic times, were ancestral to these fishes and all others. The smaller of the two fossils, one of the genus *Knightia*, is five inches long. The larger of the two is one of the genus *Diplomystus*. They swam in a sea that extended into what is now Wyoming some 50 million years ago; the specimen was collected from strata of the Green River formation at Fossil Butte, Wyo., before that fossil-rich site came under Federal protection. Part of the mineral collection of the philanthropist and amateur naturalist Boyce Thompson, the specimen was willed to the American Museum of Natural History. It forms part of the study collection of the museum's Department of Vertebrate Paleontology and was photographed through courtesy of Bobb Schaeffer of the museum's curatorial staff. The photograph was made by Fritz Goro.

Library of Congress Cataloging in Publication Data

Main entry under title:

Evolution.

"A Scientific American book."
"The nine chapters in this book originally appeared as articles in the September 1978 issue of Scientific American."
Bibliography: p.
Includes index
1. Evolution—Addresses, essays, lectures.
I. Scientific American.
QH366.2.E848 575 78-10747
ISBN 0-7167-1065-X
ISBN 0-7167-1066-8 pbk.

The nine chapters in this book originally appeared as articles in the September 1978 issue of *Scientific American*.

Printed in the United States of America

9 8 7 6 5 4 3

Contents

Foreword

These days, with work in the life sciences fired up by ideas and instrumentation borrowed from the physical sciences, the word "evolution" carries almost nostalgic connotations of the nineteenth-century, whole-animal, natural-history kind of biology. Evolution remains, nonetheless, at the very center of the work. The historic achievements in molecular biology of the last 30 years derive their significance principally from the way they illuminate our understanding of evolution. As Ernst Mayr declares in the first chapter of this book, evolution stands today as the organizing principle of biology and the general theory of life.

The contemporary kind of biology has begun to fill in the blank pages of the first 3 billion years of evolution. From resolution of the biosynthetic apparatus of the living cell, it has become possible to reconstruct the era of chemical evolution leading to the ignition of life. That was a surprisingly brief era, and it must have begun soon after the formation of the planet 4.6 billion years ago. The first fossil cells appear in rocks that are nearly 4 billion years old. We can see why it then required nearly 3 billion years to perfect the eukaryotic cell—the cell that generates and composes the tissues of all multi-celled organisms, plants as well as animals—because we have begun to appreciate the exquisite intricacy of the eukaryote's anatomy and physiology.

With the arrival of the eukaryotic cell, evolution entered the era familiar to us from textbooks. It embraces no more than one-fifth of the history of life. For nearly half of this era the reptiles abounded. On that record, the dominant dinosaur cannot be put down as a failure. The Age of Mammals began only 65 million years ago. That Age came recently to a sudden end, to be succeeded by the Age of Man.

To contemporary enterprise in the life sciences, we also owe a radical revision of our understanding of human evolution. It has been only recently established that tool-making anteceded the appearance of our species and so must have played a decisive role in the evolution of our hands and brains. Evidence from the campsites of the first toolmakers of 3 million years ago show that they were food-sharers as well. The moral evolution of our species has, therefore, been underway a long time and has proceeded along with our biological evolution.

Human and other forms of life on Earth now hang upon the rationality and humanity of man. Perhaps the understanding set out in the pages of this book can help to induce the necessary acceleration of our moral evolution.

THE EDITORS*

September, 1978

I

Evolution

Evolution

BY ERNST MAYR

Introducing a volume devoted to the history of life on the earth as it is understood in the light of the modern "synthetic" theory of evolution through natural selection, the organizing principle of biology today

The most consequential change in man's view of the world, of living nature and of himself came with the introduction, over a period of some 100 years beginning only in the 18th century, of the idea of change itself, of change over long periods of time: in a word, of evolution. Man's world view today is dominated by the knowledge that the universe, the stars, the earth and all living things have evolved through a long history that was not foreordained or programmed, a history of continual, gradual change shaped by more or less directional natural processes consistent with the laws of physics. Cosmic evolution and biological evolution have that much in common.

Yet biological evolution is fundamentally different from cosmic evolution in many ways. For one thing, it is more complicated than cosmic evolution, and the living systems that are its products are far more complex than any nonliving system; other differences will emerge in the course of this article. This *Scientific American* book deals with the origin, history and interrelations of living systems as they are understood in the light of the currently accepted general theory of life: the theory of evolution through natural selection, which was propounded more than 100 years ago by Charles Darwin, has since been modified and explicated by the science of genetics and stands today as the organizing principle of biology.

The creation myths of primitive peoples and of most religions had in common an essentially static concept of a world that, once it had been created, had not changed—and that indeed had not been in existence for very long. Bishop Ussher's 17th-century calculation that the world had been created in 4004 B.C. was noteworthy only for its misplaced precision in an age when the reach of history was still foreshortened by the limited arm's length of written records and tradition. It remained for the naturalists and philosophers of the 18th-century Enlightenment and the geologists and biologists of the 19th century to begin to extend the time dimension. In 1749 the French naturalist the Comte de Buffon first undertook to calculate the age of the earth. He reckoned it was at least 70,000 years (and suggested an age of as much as 500,000 years in his unpublished notes). Immanuel Kant was even more daring in his *Cosmogony* of 1755, in which he wrote in terms of millions or even hundreds of millions of years. Clearly both Buffon and Kant conceived of a physical universe that had evolved.

"Evolution" implies change with continuity, usually with a directional component. Biological evolution is best defined as change in the diversity and adaptation of populations of organisms. The first consistent theory of evolution was proposed in 1809 by the French naturalist and philosopher Jean Baptiste de Lamarck, who concentrated on the process of change over time: on what appeared to him to be a progression in nature from the smallest visible organisms to the most complex and most nearly perfect plants and animals and thence to man.

To explain the particular course of evolution Lamarck invoked four principles: the existence in organisms of a built-in drive toward perfection; the capacity of organisms to become adapted to "circumstances," that is, to the environment; the frequent occurrence of spontaneous generation, and the inheritance of acquired characters, or traits. The belief in the heritability of acquired characters, the error for which Lamarck is mainly remembered, was not new with him. It was a universal belief in his time, firmly grounded in folklore (one expression of which was the biblical story of Jacob and the division of the striped and speckled livestock). The belief persisted. Darwin, for example, assumed that the use or disuse of a structure by one generation would be reflected in the next generation, and so did many evolutionists until late in the century, when the German biologist August Weismann demonstrated the impossibility, or at least the improbability, of the inheritance of acquired characters. Lamarck's assumptions of a drive toward perfection and of frequent spontaneous generation were also not confirmed, but he was right in recognizing that much of evolution is what we now call adaptive. He understood, moreover, that one could explain the great diversity of living organisms only by postulating a great age for the earth, and that evolution was a gradual process.

Lamarck's main interest was evolution in the time dimension—in vertical

CHARLES DARWIN was 31 years old and had already published his journal of the round-the-world voyage of H.M.S. *Beagle* when he sat in 1840 for the watercolor portrait by George Richmond reproduced on the opposite page. By this time, judging from his notebooks, Darwin had already worked out the major features of his theory of evolution through natural selection. Recently married, he was living in London, writing a monograph on coral reefs and turning from time to time to the notes on species that were to lead in 1859 to *On the Origin of Species.*

evolution, so to speak. Darwin, in contrast, was initially intrigued by the problem of the origin of diversity, and more specifically by the origin of species through diversification in a geographical dimension—in horizontal evolution. His interest in diversification and speciation was aroused, as is well known, during his five-year voyage around the world, beginning in 1831, as naturalist on H.M.S. *Beagle*. In the Galápagos Islands, for example, he learned that each island had its own form of tortoise, of mocking bird and of finch; the various forms were closely related and yet distinctly different. Pondering his observations after his return to England, he came to the conclusion that each island population was an incipient species, and thus to the concept of the "transmutation," or evolution, of species. In 1838 he conceived of the mechanism that could account for evolution: natural selection. After more years of observation and experiment, informed by wide reading in geology, zoology and other fields, a preliminary statement of Darwin's theory of evolution through natural selection was announced in 1858 in a report to the Linnean Society of London. Alfred Russel Wallace, a young English naturalist doing fieldwork in the East Indies, had come independently to the concept of natural selection and had set down his ideas in a manuscript he mailed to Darwin; his paper was read at the meeting along with Darwin's.

Darwin's full theory, buttressed with innumerable personal observations and carefully argued, was published on November 24, 1859, in *On the Origin of Species*. His broad explanatory scheme comprised a number of component subtheories, or postulates, of which I shall single out what I take to be the four principal ones. Two of them were consistent with Lamarck's thinking. The first was the postulate that the world is not static but is evolving. Species change continually, new ones originate and others become extinct. Biotas, as reflected in the fossil record, change over time, and the older they are the more they are seen to have differed from living organisms. Wherever one looks in living nature one encounters phenomena that make no sense except in terms of evolution. Darwin's second Lamarckian concept was the postulate that the process of evolution is gradual and continuous; it does not consist of discontinuous saltations, or sudden changes.

IN ABOUT 1854, the year in which he published a large monograph on barnacles that had occupied him for some eight years, Darwin sat for this photograph. He continued meanwhile with what he called his "species work": reading, corresponding, collecting, experimenting and making notes on the subject of his major work but delaying the writing until 1856. The realization two years later that Alfred Russel Wallace had independently developed the concept of natural selection led Darwin to prepare the "abstract" we know as *On the Origin of Species.*

Darwin's two other main postulates were essentially new concepts. One was the postulate of common descent. For Lamarck each organism or group of organisms represented an independent evolutionary line, having had a beginning in spontaneous generation and having constantly striven toward perfection. Darwin postulated instead that similar organisms were related, descended from a common ancestor. All mammals, he proposed, were derived from one ancestral species; all insects had a common ancestor, and so did all the organisms of any other group. He implied, in fact, that all living organisms might be traced back to a single origin of life.

Darwin's inclusion of man in the common descent of mammals was considered by many to be an unforgivable insult to the human race, and it aroused a storm of protest. The idea of common descent had such enormous explanatory power, however, that it was almost immediately adopted by most biologists. It explained both the Linnaean hierarchy of taxonomic categories and the finding by comparative anatomists that all organisms could be assigned to a limited number of morphological types.

Darwin's fourth subtheory was that of natural selection, and it was the key to his broad scheme. Evolutionary change, said Darwin, is not the result of any mysterious Lamarckian drive, nor is it a simple matter of chance; it is the result of selection. Selection is a two-step process. The first step is the production of variation. In every generation, according to Darwin, an enormous amount of variation is generated. Darwin did not know the source of this variation, which could not be understood until after the rise of the science of genetics. All he had was his empirical knowledge of a seemingly inexhaustible reservoir of large and small differences within species.

The second step is selection through survival in the struggle for existence. In most species of animals and plants a set of parents produces thousands if not millions of offspring. Darwin's reading of Thomas Malthus told him that very few of the offspring could survive. Which ones would have the best chance of surviving? They would be those individuals that have the most appropriate combination of characters for coping with the environment, including climate, competitors and enemies; they would have the greatest chance of surviving, of reproducing and of leaving survivors, and their characters would therefore be available for the next cycle of selection.

The concept of an evolving world rather than a static one was almost universally accepted by serious scientists even before Darwin's death in 1882, and those who accepted evolution also accepted the concept of common descent (although there were those who insisted on exempting man from the common lineage). The situation was very different, however, for Darwin's two other postulates, both of which were bitterly resisted by many learned and able men for the next 50 to 80 years.

One of the postulates was the concept of gradualism. Even T. H. Huxley, who was known as "Darwin's bulldog" for his vigorous championing of most aspects of the new theory, could not accept the gradual origin of higher types and new species; he proposed a saltational origin instead. Saltationism was also popular with such biologists as Hugo De Vries, one of the rediscoverers of Gregor Mendel's laws of inheritance. He proposed a theory in 1901 according to which new species originate by mutation. As late as 1940 the geneticist Richard B. G. Goldschmidt was defending "systemic mutations" as the source of new higher types.

Three developments eventually resulted in the abandonment of such saltational theories. One development was the gradual adoption of a new attitude toward the physical world and its variation. Since the time of Plato the dominant view had been what the philosopher Karl Popper has called "essentialism": the world consisted of a limited number of unvarying essences (Plato's *eide*), of which the visible world's variable manifestations are merely incomplete and imprecise reflections. In such a view genuine change could arise only through the origin of a new essence either by creation or through a spontaneous saltation (mutation). Classes of physical objects do consist of identical entities, and physical constants are unvarying under identical conditions, and so (in the 19th century) there was no conflict between mathematics or the physical sciences and the philosophy of essentialism.

Biology required a different philosophy. Living organisms are characterized by uniqueness; every population of organisms consists of uniquely distinct individuals. In "population thinking" the mean values are the abstractions; only the variant individual has reality. The importance of the population lies in its being a pool of variations (a gene pool, in the language of genetics). Population thinking makes gradual evolution possible, and it now dominates every aspect of evolutionary theory.

The second development that led to

IN ABOUT 1880 DARWIN was photographed at Down House in Kent, where he had lived and worked since 1842. When he died in 1882 at 73, he was buried in Westminster Abbey.

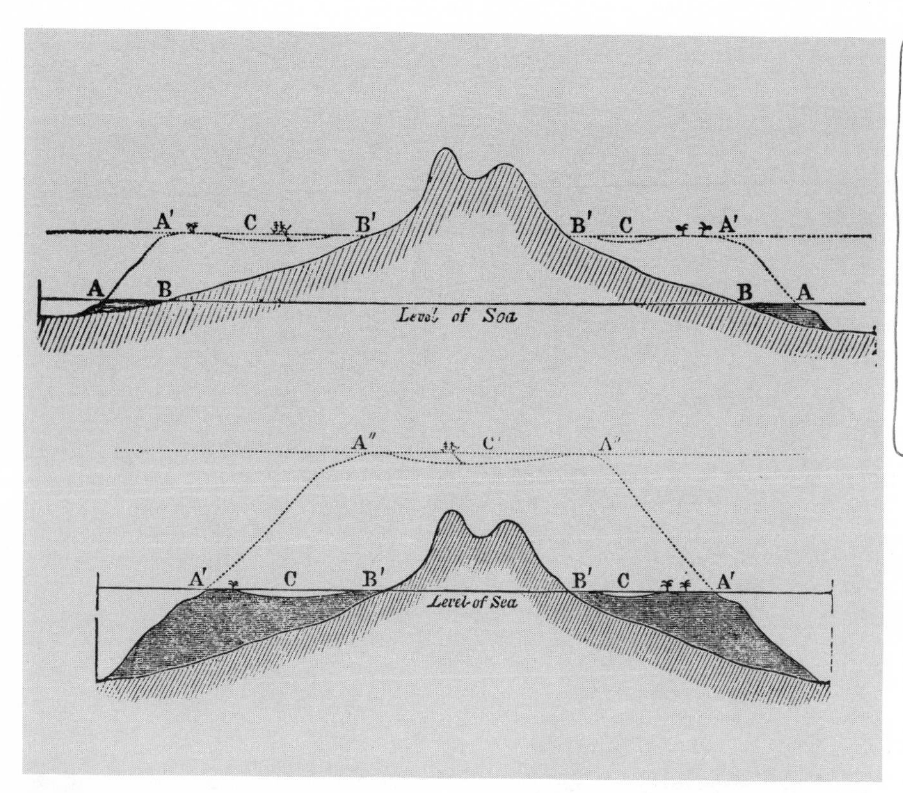

BIRTH OF AN ATOLL through subsidence of the ocean floor was illustrated by these woodcuts in Darwin's journal of the voyage of the *Beagle*. In the first stage (*top*) a fringing reef of coral (*A–B, B–A*) is built up at sea level around an island in the Pacific Ocean. As the island subsides, the coral polyps, which can survive only in shallow water, keep building the reef upward, forming a fringing reef (*A′–B′, B′–A′*) that encloses a lagoon (*C*). The island continues to subside (*bottom*) until it is below sea level; the barrier reef, growing, becomes an atoll (*A″–A″*).

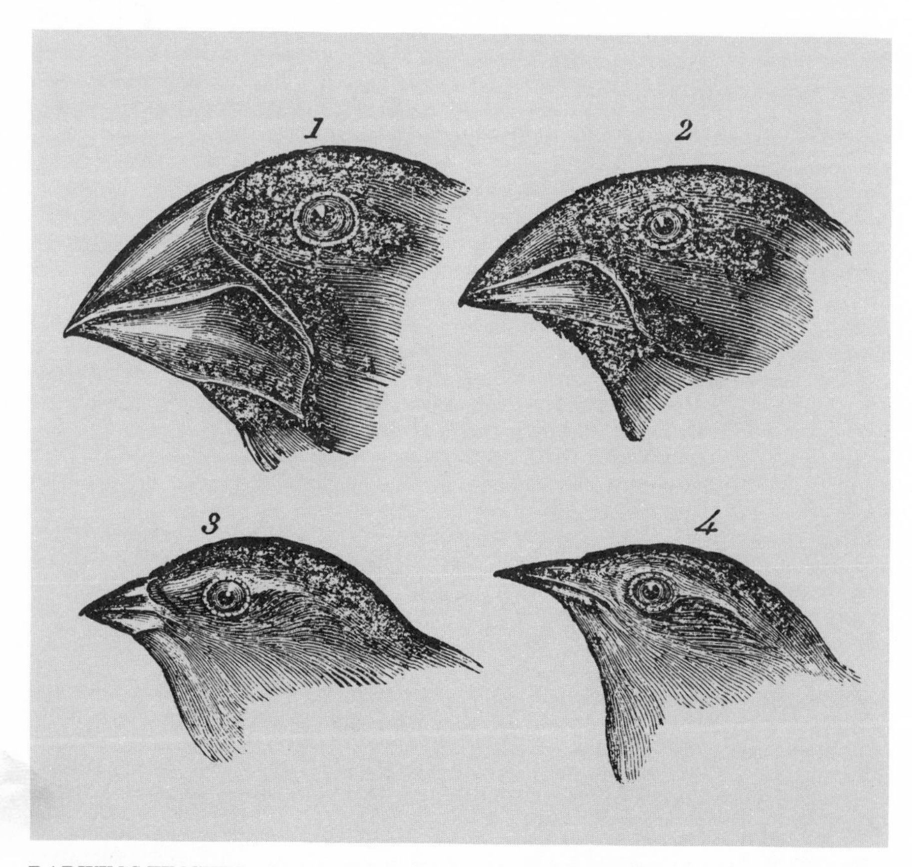

DARWIN'S FINCHES, which he observed in the Galápagos Islands and some of which were shown in this woodcut from his published journal, provided him with a major insight. Seeing the wide range of beak sizes and shapes in "one small, intimately related group of birds," he wrote, "one might really fancy that... one species had been taken and modified for different ends."

the rejection of saltation was the discovery of the immense variability of natural populations and the realization that a high variability of discontinuous genetic factors, provided there are enough of them and provided the gaps between them are sufficiently small, can manifest itself in continuous variation of the organism. The third development was the demonstration by naturalists that processes of gradual evolution are entirely capable of explaining the origin of discontinuities such as new species and new types and of evolutionary novelties such as the wings of birds and the lungs of vertebrates.

The other Darwinian concept that was long resisted by most biologists and philosophers was natural selection. At first many rejected it because it was not deterministic, and hence predictive, in the style of 19th-century science. How could a proposed "natural law" such as natural selection be entirely a matter of chance? Others attacked its "crass materialism." In the 19th century to attribute the harmony of the living world to the arbitrary workings of natural selection was to undermine the natural theologian's "argument from design," which held that the existence of a Creator could be inferred from the beautiful design of his works. Those who rejected natural selection on religious or philosophical grounds or simply because it seemed too random a process to explain evolution continued for many years to put forward alternative schemes with such names as orthogenesis, nomogenesis, aristogenesis or the "omega principle" of Teilhard de Chardin, each scheme relying on some built-in tendency or drive toward perfection or progress. All these theories were finalistic: they postulated some form of cosmic teleology, of purpose or program.

The proponents of teleological theories, for all their efforts, have been unable to find any mechanisms (except supernatural ones) that can account for their postulated finalism. The possibility that any such mechanism can exist has now been virtually ruled out by the findings of molecular biology. As the late Jacques Monod argued with particular force, the genetic material is constant; it can change only through mutation. Finalistic theories have also been refuted by the paleontological evidence, as George Gaylord Simpson has shown most clearly. When the evolutionary trend of any character—a trend toward larger body size or longer teeth, for example—is examined carefully, the trend is found not to be consistent but to change direction repeatedly and even to reverse itself occasionally. The frequency of extinction in every geological period is another powerful argument against any finalistic trend toward perfection.

As for the objection to the presumed random aspect of natural selection, it is not hard to deal with. The process is not at all a matter of pure chance. Although variations arise through random processes, those variations are sorted by the second step in the process: selection by survival, which is very much an anti-chance factor. And if it is nonetheless true that some evolution is the result of chance, it is now known that physical processes in general have a far larger probabilistic component than was recognized 100 years ago.

Even so, can natural selection explain the long evolutionary progression up to the "highest" plants and animals, including man, from the origin of life between three and four billion years ago [see "Chemical Evolution and the Origin of Life," by Richard E. Dickerson, page 30]? How can natural selection account not only for differential survival and adaptive changes within a species but also for the rise of new and differently adapted species? Again it was Darwin who suggested the right answer. An organism competes not only with other individuals of the same species but also with individuals of other species. A new adaptation or general physiological improvement will make an individual and its descendants stronger interspecific competitors and so contribute to diversification and specialization. Such specialization may often be a dead-end street, as it is in the case of adaptation to life in caves or hot springs. Many specializations, however, and particularly those that were acquired early in evolutionary history, opened up entirely new levels of adaptive radiation. These ranged from the invention of membranes and an organized cell nucleus [see "The Evolution of the Earliest Cells," by J. William Schopf, page 48] and the aggregation of cells to form multicellular organisms [see "The Evolution of Multicellular Plants and Animals," by James W. Valentine, page 66] to the advent of highly developed central nervous systems and the invention of long-continued parental care.

Evolution, as Simpson has emphasized, is recklessly opportunistic: it favors any variation that provides a competitive advantage over other members of an organism's own population or over individuals of different species. For billions of years this process has automatically fueled what we call evolutionary progress. No program controlled or directed this progression; it was the result of the spur-of-the-moment decisions of natural selection.

Darwin's uncertainty concerning the source of the genetic variability that supplies raw material for natural selection left a major hole in his argument. That hole was plugged by the science of genetics. Mendel discovered in 1865 that the factors transmitting hereditary information are discrete units transmitted by each parent to the offspring, preserved uncontaminated and reassorted in each generation. Darwin never knew of Mendel's findings, which were largely ignored until they were rediscovered in 1900.

We now know that DNA in the cell nucleus is organized in numerous self-replicating genes (Mendel's hereditary units), which can mutate to form different alleles, or alternative forms. There are structural genes that encode the information for making a specific protein and there are regulatory genes that turn the structural genes on and off. A mutated structural gene can code for a variant

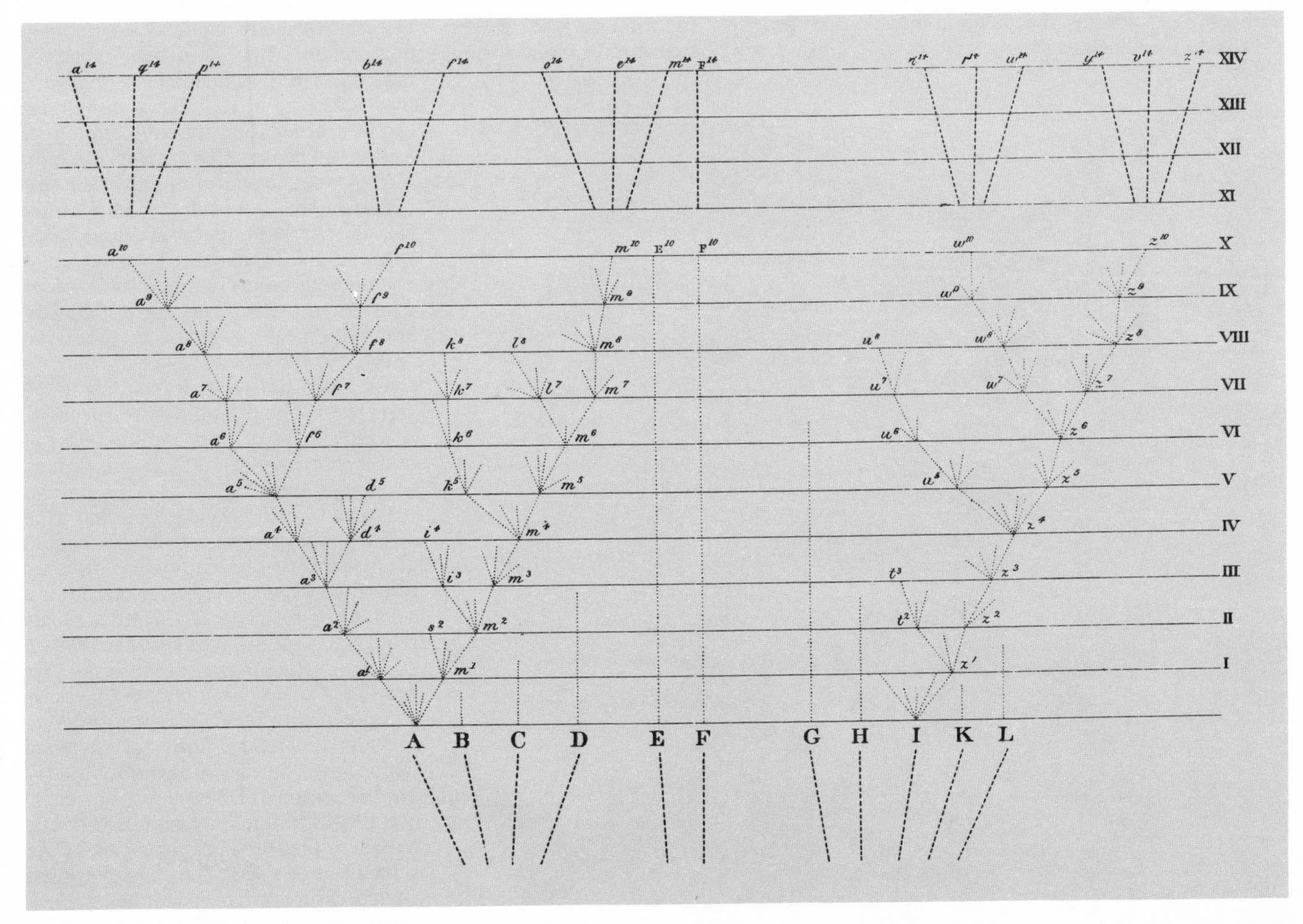

FORMATION OF NEW SPECIES through the divergence of characters and natural selection was illustrated in *On the Origin of Species*. The capital letters (*bottom*) represent species of the same genus. Horizontal lines marked by Roman numerals (*right*) represent, say, a 1,000-generation gap. Branching, diverging dotted lines represent varying offspring, the "profitable" ones of which are "preserved or naturally selected." Some species (*B, C and so on*) die out; some (*E, F*) remain essentially unchanged. Some (*A, I*) diverge widely, giving rise after many generations to new varieties (a^1, m^1, z^1) that diverge in turn, giving rise to increasingly divergent varieties that eventually become distinct new species (a^{14}, q^{14}, p^{14} *and so on*). After longer intervals these may become new genera or even higher categories.

protein, leading to a variant character. The genes are arrayed on chromosomes and may recombine with one another during meiosis, the cellular process that precedes the formation of germ cells in sexually reproducing species. The diversity of genotypes (full sets of genes) that can be produced during meiosis is almost unimaginably great, and much of that diversity is preserved in populations in spite of natural selection [see "The Mechanisms of Evolution," by Francisco J. Ayala, page 14].

Strangely, the early Mendelians did not accept the theory of natural selection. They were essentialists and saltationists, and they looked on mutation as the probable driving force in evolution. That began to change with the development of population genetics in the 1920's. Eventually, during the 1930's and 1940's, a synthesis was achieved, expressed in and largely brought about by books written by Theodosius Dobzhansky, Julian Huxley, Bernhard Rensch, Simpson, G. Ledyard Stebbins and me. The new "synthetic theory" of evolution amplified Darwin's theory in the light of the chromosome theory of heredity, population genetics, the biological concept of a species and many other concepts of biology and paleontology. The new synthesis is characterized by the complete rejection of the inheritance of acquired characters, an emphasis on the gradualness of evolution, the realization that evolutionary phenomena are population phenomena and a reaffirmation of the overwhelming importance of natural selection.

The understanding of the evolutionary process achieved by the synthetic theory has had a profound effect on all biology. It led to the realization that every biological problem poses an evolutionary question, that it is legitimate to ask with respect to any biological structure, function or process: Why is it there? What was its selective advantage when it was acquired? Such questions have had an enormous impact on every area of biology, notably molecular biology, behavioral studies and ecology [see "The Evolution of Ecological Systems," by Robert M. May, page 80].

Philosophers and physical scientists as well as lay people continue to have trouble understanding the modern theory of organic evolution through natural selection. At the risk of repeating some points I have already made in a historical context, let me outline the special features of the current theory, in particular drawing attention to what distinguishes organic evolution from cosmic evolution and other processes dealt with by physical scientists.

Evolution through natural selection is (I repeat!) a two-step process. The first step is the production (through recombination, mutation and chance events) of genetic variability; the second is the ordering of that variability by selection. Most of the variation produced by the first step is random in that it is not caused by, and is unrelated to, the current needs of the organism or the nature of its environment.

Natural selection can operate successfully because of the inexhaustible supply of variation made available to it owing to the high degree of individuality of biological systems. No two cells within an organism are precisely identical; each individual is unique, each species is unique and each ecosystem is unique. Many nonbiologists find the extent of organic variability incomprehensible. It is totally incompatible with traditional essentialist thinking and calls for a very different conceptual framework: population thinking. (The individuality of biological systems and the fact that there are multiple solutions for almost any environmental problem combine to make organic evolution nonrepeatable. Deterministically inclined astronomers are convinced by statistical reasoning that what has happened on the earth must also have happened on planets of stars other than the sun. Biologists, impressed by the inherent improbability of every single step that led to the evolution of man, consider what Simpson called "the prevalence of humanoids" exceedingly improbable.)

Uniquely different individuals are organized into interbreeding populations and into species. All the members are "parts" of the species, since they are derived from and contribute to a single gene pool. The population or species as a whole is itself the "individual" that undergoes evolution; it is not a class with members.

Every biological individual has a peculiarly dualistic nature. It consists of a genotype (its full complement of genes, not all of which may be expressed) and a phenotype (the organism that results from the translation of genes in the genotype). The genotype is part of the gene pool of the population; the phenotype competes with other phenotypes for reproductive success. This success (which defines the "fitness" of the individual) is not determined intrinsically but is the result of multiple interactions with enemies, competitors, pathogens and other selection pressures. The constellation of such pressures changes with the seasons, through the years and geographically.

The second step of natural selection, selection itself, is an extrinsic ordering principle. In a population of thousands or millions of unique individuals

JEAN BAPTISTE DE LAMARCK, the French naturalist and philosopher who was the first consistent evolutionist, understood that the earth is very old, that evolution is gradual and that organisms adapt. Lamarck also believed, however, in the inheritance of acquired characters.

some will have sets of genes that are better suited to the currently prevailing assortment of ecological pressures. Such individuals will have a statistically greater probability of surviving and of leaving survivors than other members of the population. It is this second step in natural selection that determines evolutionary direction, increasing the frequency of genes and constellations of genes that are adaptive at a given time and place, increasing fitness, promoting specialization and giving rise to adaptive radiation and to what may be loosely described as evolutionary progress [see "Adaptation," by Richard C. Lewontin, page 114].

Selectionist evolution, in other words, is neither a chance phenomenon nor a deterministic phenomenon but a two-step tandem process combining the advantages of both. As the pioneering population geneticist Sewall Wright wrote: "The Darwinian process of continued interplay of a random and a selective process is not intermediate between pure chance and pure determinism, but in its consequences qualitatively utterly different from either."

No Darwinian I know questions the fact that the processes of organic evolution are consistent with the laws of the physical sciences, but it makes no sense to say that biological evolution has been "reduced" to physical laws. Biological evolution is the result of specific processes that impinge on specific systems, the explanation of which is meaningful only at the level of complexity of those processes and those systems. And the classical theory of evolution has not been reduced to a "molecular theory of evolution," an assertion based on such reductionist definitions of evolution as "a change in gene frequencies in natural populations." This reductionist definition omits the crucial aspects of evolution: changes in diversity and adaptation. (Once I gave a lump of sugar to a raccoon in a zoo. He ran with it to his water basin and washed it vigorously until there was nothing left of it. No complex system should be taken apart to the extent that nothing of significance is left.)

After the new synthesis of the 1930's and 1940's was achieved a few nonevolutionists asked whether it did not mark the end of research in evolution, whether all the questions had not been answered. The answer to both questions is decidedly no, as is made clear by the exponential increase in the number of publications in evolutionary biology. Let me mention some problems that currently interest workers in the field.

One major subject of inquiry is the role of chance. As far back as 1871 it was proposed that perhaps only some evolutionary change is due to selection, with much or even most change being

ALFRED RUSSEL WALLACE, as a young naturalist working in the East Indies, independently developed a theory of natural selection; his paper on the subject was read along with Darwin's in 1858. Later he differed with Darwin about the mechanisms of human evolution: Wallace believed that natural selection alone could not account for man's higher capacities.

T. H. HUXLEY, distinguished for brilliant work in many areas of biology, took on himself the role of Darwin's "general agent" and "bulldog," explicating and praising *On the Origin of Species* in a book review published in *The Times* of London and in many articles and lectures.

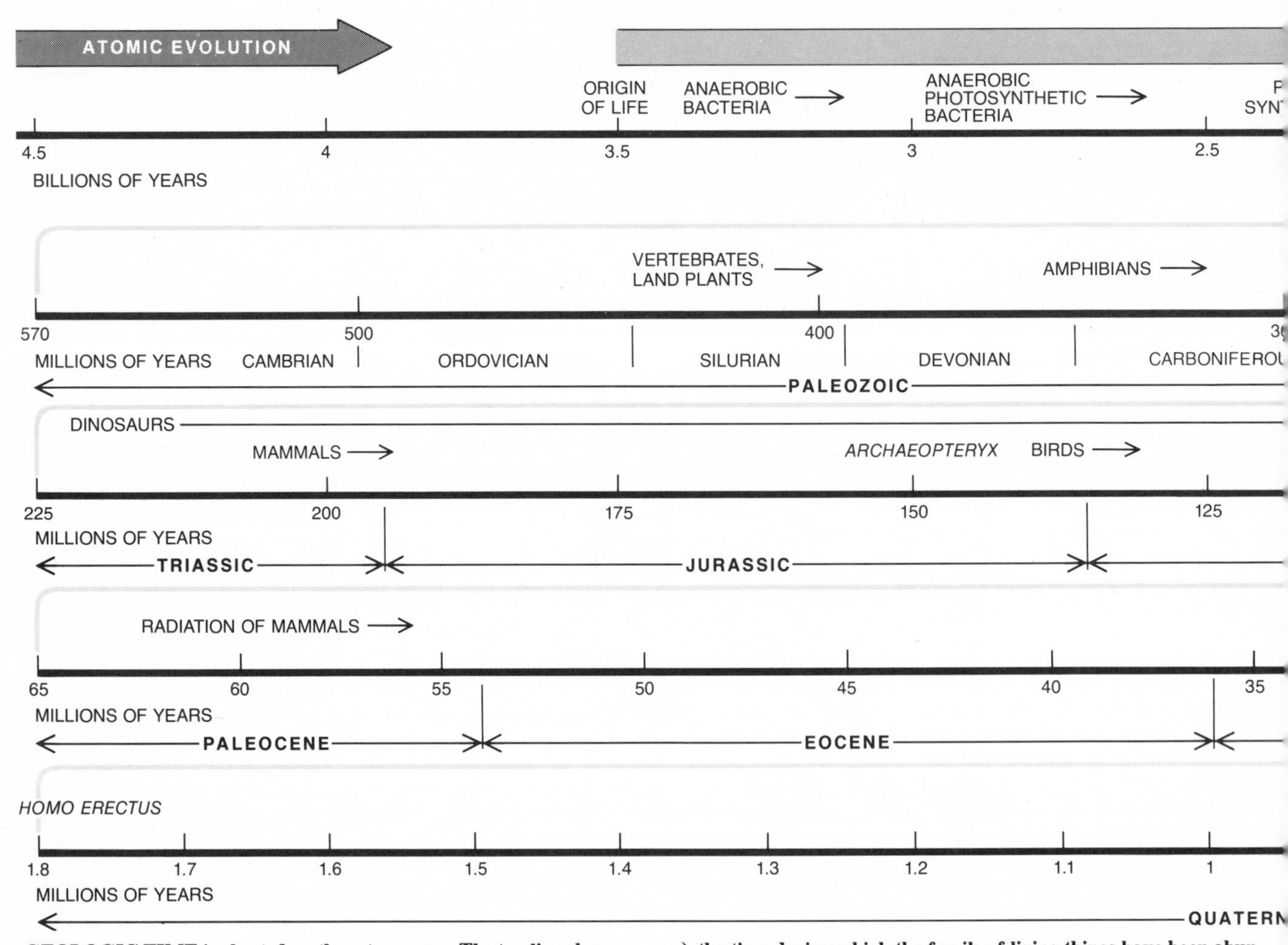

GEOLOGIC TIME is charted on these two pages. The top line shows the full sweep from the origin of the earth some 4.6 billion years ago to the present day. The relatively short span of Phanerozoic time (*color*), the time during which the fossils of living things have been abundant in the geological record, is enlarged in the second line from the top, and successively shorter periods (*color*) are enlarged in the next

due to accidental variation, or to what are now called "neutral" mutations; the suggestion has been repeated many times since then. The problem acquired a new dimension when the technique of electrophoresis made it possible to detect small differences in the composition of a particular enzyme in a large random sample of individuals, thereby revealing the enormous extent of allelic variability. What part of that variability is evolutionary "noise" and what part is due to selection? How can one partition the variability into neutral and into relatively significant alleles?

The discovery of molecular biology that there are regulatory genes as well as structural ones poses new evolutionary questions. Is the rate of evolution of the two kinds of genes the same? Are they equally susceptible to natural selection? Is one kind of gene more important than the other in speciation or in the origin of higher taxa? (For example, the structural genes of the chimpanzee and of man appear to be remarkably similar. Is it perhaps the regulatory genes that make for most of the difference between us and them?) Are there still other kinds of genes?

Darwin's favorite problem, that of the multiplication of species, has again become a focus of research. In certain groups of organisms, such as birds, new species seem to originate exclusively by geographical speciation: through the genetic restructuring of populations isolated from the remainder of a species' range, as on an island. In plants and in a few groups of animals, however, a different form of speciation can be effected through polyploidy, the doubling of the set of chromosomes, because polyploid individuals are immediately isolated reproductively from their parents. Another mode of speciation is "sympatric" speciation in parasites or in insects that are adapted to life on a specific host plant. Occasionally a new host species is colonized accidentally, and the descendants of the immigrant, perhaps aided by having favorable genes, come to constitute a well-established colony. In such a case there will be strong selection of genes that favor reproduction with other individuals living on the new host species, so that conditions may favor the development of a new race adapted to the new host, and eventually of a new host-specific species. The frequency of sympatric speciation is still a matter of controversy. The respective role of genes and chromosomes in speciation is yet another controversial area.

In few areas of biology has the introduction of evolutionary thinking been as productive as it has in behavioral biology. The classical ethologists showed that such behavior patterns as the signaling displays of courtship can be as indicative of taxonomic relations as structural characters are. Classifications based on behavior have been worked out that agree remarkably well with systems based on structure, and the behavioral data have often provided decisive clues where the morphological evidence was ambiguous. More important has been the demonstration that behavior often—perhaps invariably—serves as a pacemaker in evolution. A change in behavior, such as the selection of a new habitat or food source, sets up new selective pressures and may lead to important

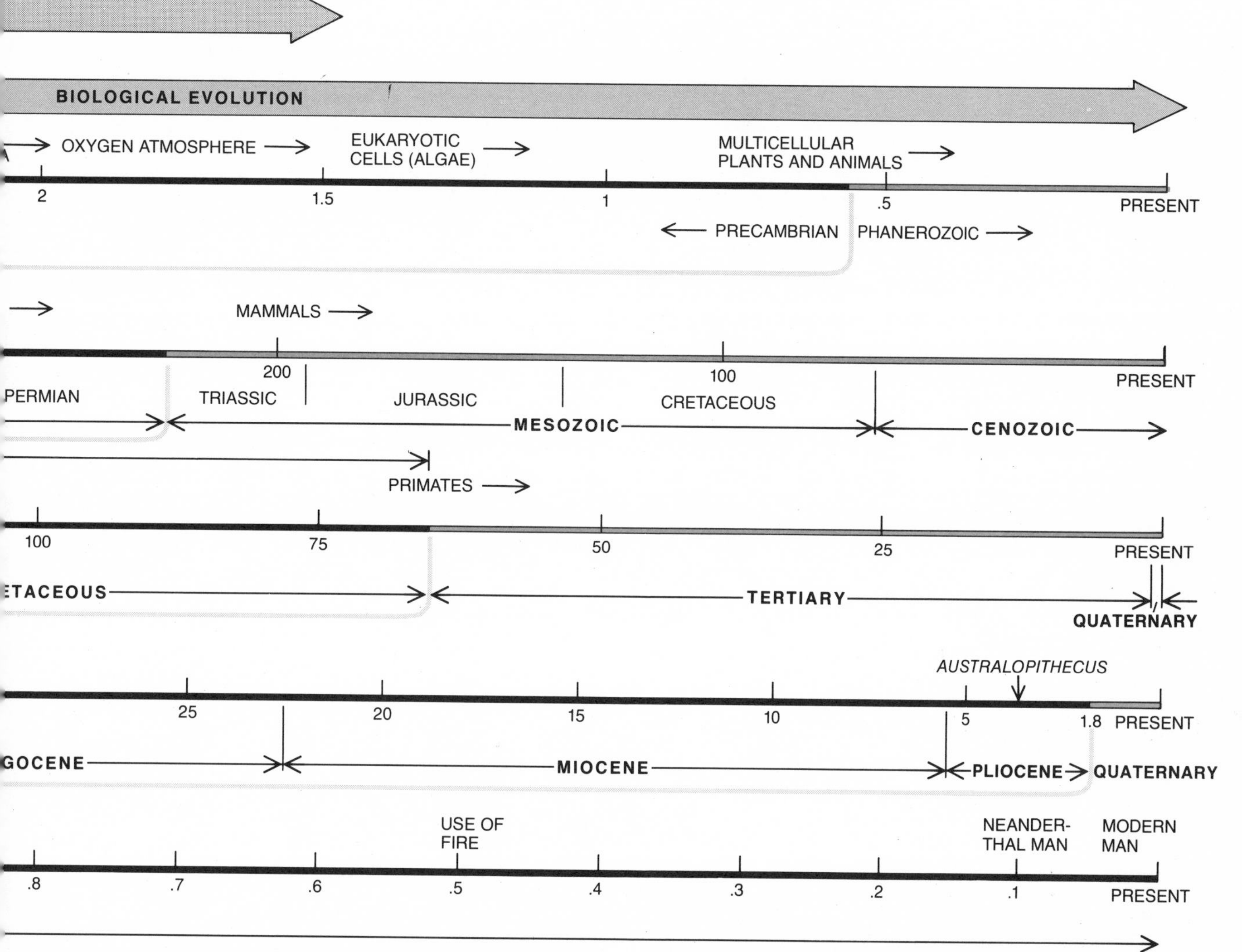

three lines. Three stages of evolution are shown at the top, with biological evolution beginning some 3.5 billion years ago with the appearance of the first living cells known. The three eras of Phanerozoic time (Paleozoic, Mesozoic and Cenozoic) are divided in turn into 11 periods; the Tertiary period is divided into five epochs, and the Quaternary period comprises the Pleistocene epoch and recent time.

adaptive shifts. There is little doubt that some of the most important events in the history of life, such as the conquest of land or of the air, were initiated by shifts in behavior. The selection pressures that potentiate such evolutionary progress are now receiving special attention [see "The Evolution of Behavior," by John Maynard Smith, page 92].

The perception that the world is not static but forever changing and that our own species is the product of evolution has inevitably had a fundamental impact on human understanding. We now know that the evolutionary line to which we belong arose from apelike ancestors over the course of millions of years, with the crucial steps having taken place during the past million years or so [see "The Evolution of Man," by Sherwood L. Washburn, page 104]. We know that natural selection must have been responsible for this advance. What do past events enable one to predict with regard to the future of mankind? Since there is no finalistic element in organic evolution and no inheritance of acquired characters, selection is obviously the only mechanism potentially capable of influencing human biological evolution.

That conclusion poses a dilemma. Eugenics, or deliberate selection, would be in conflict with cherished human values. Even if there were no moral objections, the necessary information on which to base such selection is simply not yet available. We know next to nothing about the genetic component of nonphysical human traits. There are innumerable and very different kinds of "good," "useful" or adapted human beings. Even if we could select a set of momentarily ideal characteristics, the changes generated in society by technological advances come so rapidly that no one could predict what particular blend of talents would lead in the future to the most harmonious human society. "Mankind is still evolving," Dobzhansky said, but we cannot know where it is headed biologically.

There is another kind of evolution, however: cultural evolution. It is a uniquely human process by which man to some extent shapes and adapts to his environment. (Whereas birds, bats and insects became fliers by evolving genetically for millions of years, Dobzhansky pointed out, "man has become the most powerful flier of all, by constructing flying machines, not by reconstructing his genotype.") Cultural evolution is a much more rapid process than biological evolution. One of its aspects is the fundamental (and oddly Lamarckian) ability of human beings to evolve culturally through the transmission from generation to generation of learned information, including moral—and immoral—values. Surely in this area great advances can still be made, considering the modest level of moral values in mankind today. Even though we have no way of influencing our own biological evolution, we can surely influence our cultural and moral evolution. To do so in directions that are adaptive for all mankind would be a realistic evolutionary objective, but the fact remains that there are limits to cultural and moral evolution in a genetically unmanaged human species.

II

The Mechanism of Evolution

The Mechanisms of Evolution

BY FRANCISCO J. AYALA

The rapid advances of molecular genetics over the past two decades have accounted for the origin of mutations and have revealed that the variation within species is much greater than Darwin postulated

In the 119 years since the publication of *On the Origin of Species* Darwin's basic principles have been progressively refined. According to Darwin, the basis of evolution is the occurrence of random heritable modifications in the individuals of a population. The advantageous modifications are then adopted and the disadvantageous ones are discarded through natural selection: the differential survival and reproduction of genetically variant individuals. In this way evolutionary adaptation involves a mixture of variation and selection, of chance and necessity.

Darwin thought of variation as a transient phenomenon. Because a population of organisms is closely adapted to its environment, he argued, the vast majority of modifications will be disadvantageous and the modified individuals will accordingly be eliminated by natural selection. In the rare event that a modification is advantageous it will render the individual more likely to survive and reproduce. As a result the advantageous modification will gradually spread to all the members of the population over the generations, ultimately replacing the type that was formerly dominant.

Darwin's theory implies that natural populations are made up of a more or less common genetic type with a few rare variants. In recent years this assumption has been contradicted by evidence that natural populations possess an enormous reservoir of genetic variation, suggesting that the role of chance in the evolutionary process is subtler than Darwin supposed. The advances in molecular biology, together with the statistical approach to evolution provided by population genetics, have enabled biologists to better understand where genetic variation comes from, how it is maintained in populations and how it contributes to evolutionary change.

In Darwin's day the science of genetics had not yet been born. The discrete units of heredity called genes were first identified by Gregor Mendel in Darwin's lifetime but did not become widely known until the 20th century. Darwin's vague but prescient notion of random fluctuations in the hereditary material nonetheless turned out to be an approximation of Mendel's more precise concept of genetic variation, and so Mendelian genetics could be incorporated into the theory of natural selection without too much difficulty. The fusion of the two disciplines from the early 1920's through the late 1950's is often referred to as Neo-Darwinism or the modern synthesis.

The dramatic discoveries of molecular genetics over the past 20 years have led to yet another synthesis, encompassing an understanding of evolutionary processes at the molecular level. A gene is now known to be a segment of one of the extremely long DNA molecules in the cell that store the organism's genetic information in their structure. The sequence of four kinds of nucleotide base (adenine, cytosine, guanine and thymine) along each strand of the DNA double helix represents a linear code. The information contained in that code directs the synthesis of specific proteins; the development of an organism depends on the particular proteins it manufactures. Proteins are made up of long chains of amino acids, and the specific properties of each protein are determined by the sequence of amino acids in the chain. This sequence is in turn specified by the sequence of nucleotide bases in the DNA of the genes.

The genetic information stored in the DNA molecule is expressed in two steps. In the first process, called transcription, the sequence of nucleotide bases along one of the DNA strands is copied onto a complementary strand of RNA (which is made up of the same nucleotide bases as DNA except that thymine is replaced by the closely related uracil). In the second process, called translation, the genetic program of the organism is "read" from the RNA in codons, or successive groups of three nucleotide bases. The four RNA bases form 64 different codons that specify the 20 common amino acids in proteins. (The discrepancy between the 64 codons and the 20 amino acids is due to the redundancy of the genetic code and the fact that certain codons represent instructions such as "Start" and "Stop.")

In protein synthesis the amino acids specified by the sequence of codons along the gene are added one by one to the growing chain. Once the protein has been assembled it spontaneously assumes a specific three-dimensional form and begins to function as an enzyme, as a structural component or in some other biological role. The characteristics and behavior of organisms depend ultimately on the sequences of amino acids in

GENETIC VARIATION WITHIN A SPECIES is apparent in the color patterns on the elytra (wing covers) of the Asiatic lady beetle *Harmonia axyridis*, as is illustrated in the painting on the opposite page. A species indigenous to Siberia, Japan, Korea and China, *H. axyridis* occurs in a number of discrete variant forms with different geographical distributions. Variant *19-signata* (*top three rows*) has many patterns of black spots on a yellow field and even a few solid-black individuals, variant *aulica* (*fourth row*) has a large pair of yellow spots on a black field, variant *axyridis* (*fifth row*) has spots that may range in color from orange-yellow to pale orange and variant *spectabilis* (*sixth row*) has red spots on a black field. The geographical distribution of the populations of this species is quite sharp: west-central Siberia is occupied by a population nearly uniform for the black-background *axyridis* pattern. Farther eastward the populations become more variable, with the yellow-background forms such as *signata* increasing in frequency. The red-on-black *spectabilis* pattern is found only in the Far East. The diverse color patterns are believed to be determined by a series of variant forms of the same gene. Although a discrete and striking variation of this type, called a polymorphism, is rare, subtler types of variation are seen in all living species, including man. In addition natural populations harbor large reservoirs of hidden variation, enabling them to adapt to changing environments.

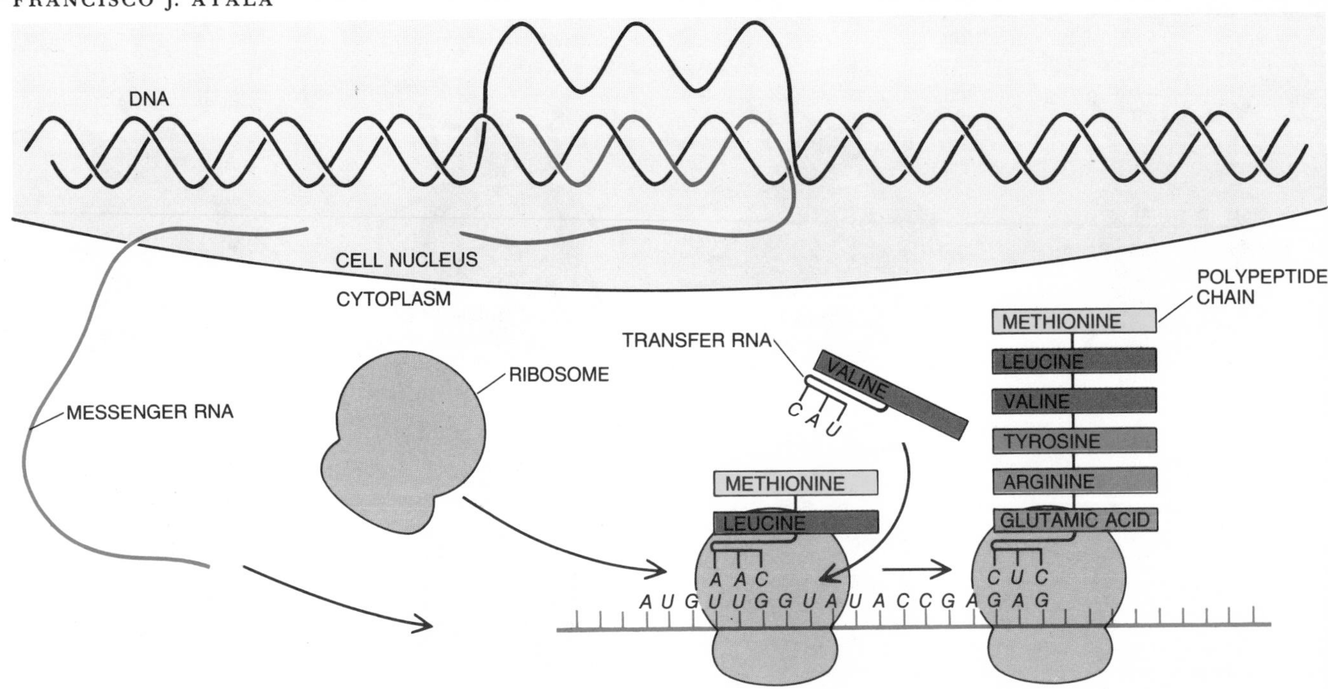

"CENTRAL DOGMA" of molecular genetics states that genetic information flows from DNA to messenger RNA to protein. Genes are relatively short segments of the long DNA molecules in cells. The DNA molecule comprises a linear code made up of four types of nucleotide base: adenine (*A*), cytosine (*C*), guanine (*G*) and thymine (*T*). The code is expressed in two steps: first the sequence of nucleotide bases in one strand of the DNA double helix is transcribed onto a single complementary strand of messenger RNA (which has the same bases as DNA except that thymine is replaced by the closely related uracil, or *U*). The messenger RNA is then translated into protein by means of complementary transfer-RNA molecules, which add amino acids one by one to the growing chain as the ribosome moves along the messenger-RNA strand. Each of the 20 amino acids found in proteins is specified by a "codon" made up of three sequential RNA bases.

their proteins, and evolution consists largely in the progressive substitution of one amino acid for another.

The new understanding of the chemical nature of the gene has provided a view of mutation at the molecular level. A mutation can be considered an error in the replication of DNA prior to its translation into protein. Such an error is often confined to the replacement of one nucleotide-base pair by another (a point mutation), and it may lead to the replacement of one amino acid by another in the protein specified for by that gene. Point mutations that result in the substitution of an amino acid are called missense mutations; those that convert the codon for an amino acid into a "stop" codon are called nonsense mutations. Other mutations may involve the insertion of a nucleotide into the DNA mole-

FIRST RNA NUCLEOTIDE BASE	SECOND RNA NUCLEOTIDE BASE: *U*	*C*	*A*	*G*	THIRD RNA NUCLEOTIDE BASE
URACIL (*U*)	PHENYLALANINE	SERINE	TYROSINE	CYSTEINE	*U*
	PHENYLALANINE	SERINE	TYROSINE	CYSTEINE	*C*
	LEUCINE	SERINE	STOP	STOP	*A*
	LEUCINE	SERINE	STOP	TRYPTOPHAN	*G*
CYTOSINE (*C*)	LEUCINE	PROLINE	HISTIDINE	ARGININE	*U*
	LEUCINE	PROLINE	HISTIDINE	ARGININE	*C*
	LEUCINE	PROLINE	GLUTAMINE	ARGININE	*A*
	LEUCINE	PROLINE	GLUTAMINE	ARGININE	*G*
ADENINE (*A*)	ISOLEUCINE	THREONINE	ASPARAGINE	SERINE	*U*
	ISOLEUCINE	THREONINE	ASPARAGINE	SERINE	*C*
	ISOLEUCINE	THREONINE	LYSINE	ARGININE	*A*
	START/METHIONINE	THREONINE	LYSINE	ARGININE	*G*
GUANINE (*G*)	VALINE	ALANINE	ASPARTIC ACID	GLYCINE	*U*
	VALINE	ALANINE	ASPARTIC ACID	GLYCINE	*C*
	VALINE	ALANINE	GLUTAMIC ACID	GLYCINE	*A*
	VALINE	ALANINE	GLUTAMIC ACID	GLYCINE	*G*

NEUTRAL · AROMATIC · BASIC · ACIDIC · SULFUR-CONTAINING

DICTIONARY OF THE GENETIC CODE is tabulated here in the language of messenger RNA. The code is universal: all organisms, from the lowliest bacterium to man, use the same set of RNA codons to specify the same 20 amino acids. In addition *AUG* serves as a "start" codon to signal the beginning of the messenger-RNA transcript, and *UAA, UAG* and *UGA* serve as "stop" codons that signal the end of the transcript and cause the completed protein to be released from the ribosome. The code is highly redundant in that several codons specify the same amino acid. Nevertheless, certain point mutations (single substitutions of one nucleotide-base pair for another in the DNA molecule) may change a codon so that it specifies a different amino acid. In the table amino acids with similar chemical properties have been grouped together by color. Point mutations that result in the substitution of one amino acid for another of the same group ("conservative" mutations) usually lead to subtle changes in the structure and function of a protein. In contrast, point mutations that result in the substitution of one amino acid for another of a different group may lead to drastic changes in the protein. Because of the clustering of amino acids of similar type most point mutations lead to conservative substitutions and hence to minor changes in proteins.

cule or the deletion of a nucleotide from it; such mutations may have more pervasive effects by shifting the "frame" in which the nucleotide sequence is read, and they may lead to several missense or nonsense substitutions. If these DNA mutations occur in the germ cells of the organism, they will be passed on to the next generation.

In addition to changes in the structure of the genes by mutation, evolution involves changes in the amount and organization of the genes. A human being has in each cell many times more DNA than our single-cell ancestors of a billion years ago had. Evolutionary increments (or decrements) in the hereditary material occur largely by means of duplications (or deletions) of DNA segments; the duplicated segments can then evolve toward serving new functions while the preexisting segments retain the original function.

The forces that give rise to gene mutations operate at random in the sense that genetic mutations occur without reference to their future adaptiveness in the environment. In other words, a mutant individual is no more likely to appear in an environment in which it would be favored than in one in which it would be selected against. If a favored mutation does appear, it can be viewed as exhibiting a "preadaptation" to that particular environment: it did not arise as an adaptive response but rather proved to be adaptive after it appeared.

A population consisting of several million individuals is likely to have a few mutations per generation in virtually every gene carried by the population. Mutations that give rise to substantial changes in the physical characteristics of the organism, however, are unlikely to be advantageous. Since a population is usually well adapted to its environment, major changes are usually maladaptive, just as a large random change in the construction of a clock (the removal of a spring or the addition of a gear) is not likely to improve its functioning. Most evolutionary changes seem to occur by the gradual accumulation of minor mutations (analogous to the tightening of a screw) accompanied by slow transitions in the physical characteristics of individuals in the population.

The DNA molecules in the nucleus of higher cells are associated with protein and packed into the dense bodies called chromosomes. The number of chromosomes in the cell nucleus differs from species to species: eight in the fruit fly *Drosophila,* 20 in corn, 24 in the tomato, 40 in the house mouse, 46 in man, 48 in the potato. A substantial reorganization of the hereditary material can result from transpositions of chromosomal segments, each of which comprises hundreds or thousands of nucleotide bases. The total number of chromosomes can be increased by duplication

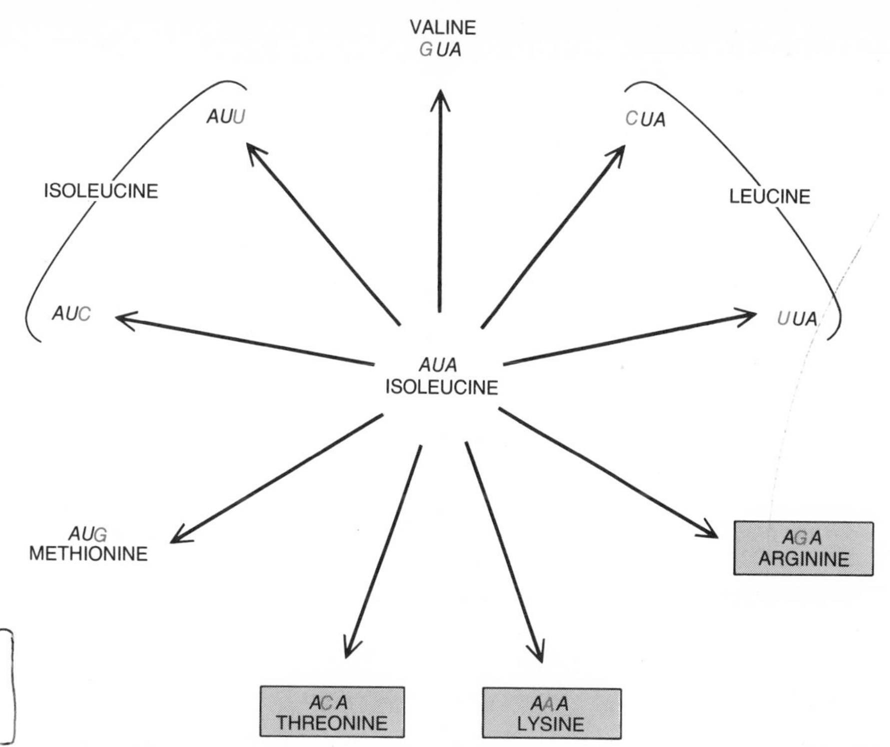

POINT MUTATIONS occur randomly during the replication of DNA molecules. They can be induced by ionizing radiation, elevated temperatures and a variety of chemical reagents or can arise naturally through other processes. This diagram shows that substitutions at the first, second or third position in the messenger-RNA codon for the amino acid isoleucine can give rise to nine new codons that code for a total of six different amino acids. (Because of redundancy of the genetic code some point mutations cause no change in amino acid.) Codons in boxes specify amino acids with chemical properties that differ sharply from those of isoleucine.

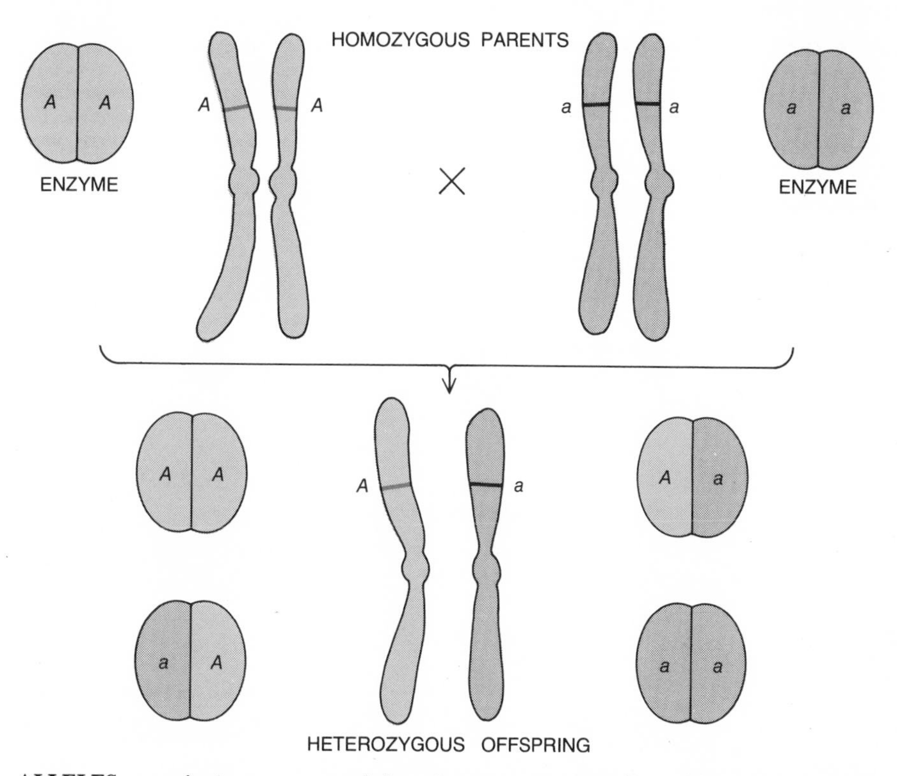

ALLELES, or variant genes, are carried on chromosomes at specific positions termed loci. In this diagram the individual at the left has allele *A* at a given locus on two homologous, or corresponding, chromosomes. The individual at the right has a different allele *a* at the same locus on two homologous chromosomes. Because these individuals possess two copies of the same allele, they are called homozygotes. When they are crossed, their offspring will possess one copy of each allele, making them heterozygotes. Because each allele codes for a slightly different protein, heterozygosity can be inferred from the presence of two variants of a given protein in a single individual. For example, here the enzyme coded for by the locus is made up of two identical protein chains that combine spontaneously. Each homozygote will manufacture either the *AA* or the *aa* form of the enzyme, whereas the heterozygote will manufacture *AA, aa* and *Aa.*

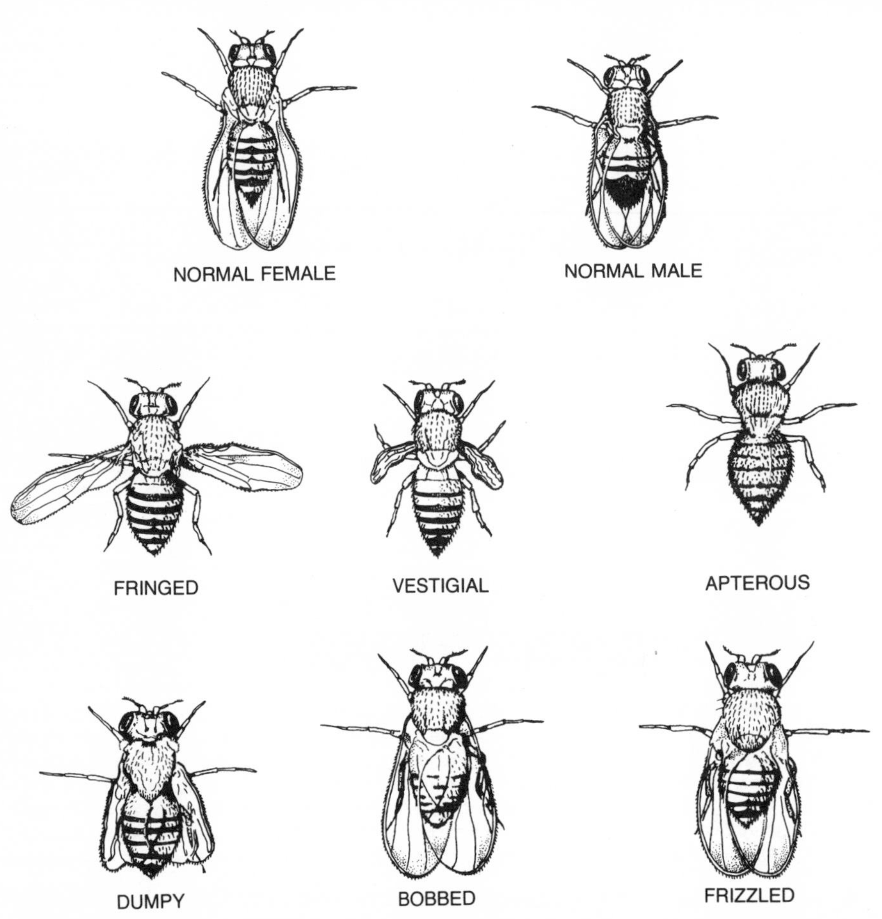

DISADVANTAGEOUS RECESSIVE ALLELES in a population of the fruit fly *Drosophila melanogaster* gave rise to the gross anatomical defects shown here. These disadvantageous alleles (originally created in the laboratory by ionizing radiation) are expressed only when they are homozygous; in the heterozygous state they are usually concealed. Their existence was revealed by inbreeding closely related individuals so that many disadvantageous alleles became homozygous in the progeny and were therefore expressed. Such alleles are maintained in a population at low frequencies; they may become advantageous when the environment changes.

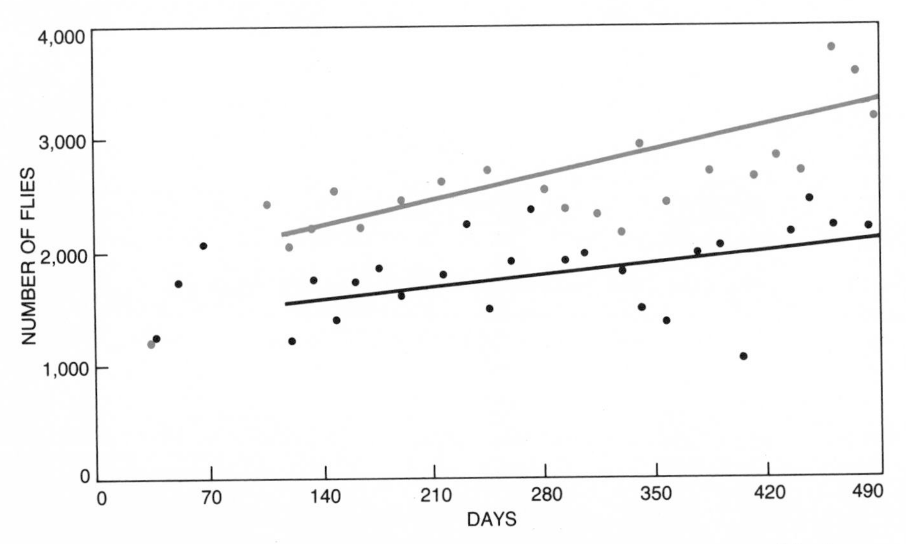

EFFECT OF GENETIC VARIATION on the rate of evolution was demonstrated by the author in experiments performed on *Drosophila serrata*. Two populations of the species were studied: one derived from a single strain and the other derived from crossbreeding two different strains, so that it had about twice as much genetic variation. Both types of population were then placed in closed bottles for 25 generations under conditions of intense competition for food and living space, which fosters rapid evolutionary change. Although both the single-strain and the mixed-strain populations became increasingly adapted to the laboratory conditions, as evidenced by a rise in population over a period of time, the average rate of increase in the mixed population (*color*) was about twice that of the single-strain population (*black*). The greater the variation stored in a population is, the more readily it is able to adapt to a new environment.

or reduced by fusion. A segment of a chromosome can be deleted, an extra piece can be inserted or a segment can be removed, inverted and put back. A segment from one chromosome can be transferred to another, or noncorresponding pieces can be exchanged. All these chromosomal aberrations alter the organization of the genes and contribute new raw material for evolutionary change.

Of the 46 chromosomes in every human cell, 23 are copies of those originating in the sperm of the father and the other 23 are copies of those originating in the egg of the mother. The genes thus occur in pairs, one on a maternal chromosome and the other on the homologous, or corresponding, paternal chromosome. The two genes in a pair are said to occupy a certain locus, or position, on each of the homologous chromosomes. For example, there is a locus on one pair of homologous chromosomes that codes for eye color. Each chromosome may comprise many thousands of gene loci.

A gene at a given locus may have variant forms known as alleles. There may be several alleles at a locus in a large population, although there can be only two in any one individual. Each allele arises by mutation from a preexisting gene and may differ from it at one or several parts of its nucleotide-base sequence. When the two alleles at a certain locus on the homologous chromosomes in an individual are identical, the individual is said to be homozygous at that locus; when the two alleles are different, the individual is said to be heterozygous at that locus.

Hereditary variation, as reflected in the existence of multiple alleles in a population, is clearly a prerequisite for evolutionary change. If all the individuals in a population are homozygous for the same allele at a given locus, there can be no evolution at that locus until a new allele arises by mutation. If, on the other hand, two or more alleles are present in a population, the frequency of one allele can increase at the expense of the other or others as a consequence of natural selection. Of course, the selective value of an allele is not fixed. The environment is variable in space and time; under certain conditions one allele is favored and under different conditions another allele is favored. Hence a population that has considerable amounts of genetic variation may be hedged against future changes in the environment.

Laboratory experiments have demonstrated that the greater the amount of genetic variation in a population, the faster its rate of evolution. In one experiment two populations of the fruit fly *Drosophila* were bred so that one population had initially about twice as much genetic variation as the other. The populations were then allowed to evolve in the laboratory for 25 generations with intense competition for food and living space, conditions that tend to stimulate rapid evolutionary change. Although both types of population evolved, gradually becoming better adapted to the laboratory environment, the rate of evolution was substantially higher in the population that had the greater initial variation.

The question of how much variation exists in natural populations is therefore of central interest to biologists, since it determines to a large extent the evolutionary plasticity of a species. The task of estimating genetic variation, however, is a difficult one since much genetic variation is hidden in each generation and is not expressed as manifest traits. The reason is that at a given locus in a heterozygous individual one allele is usually dominant and the other is recessive, that is, only the dominant allele is expressed in the heterozygous state. If a human being has a dominant allele for brown eye color and a recessive one for blue eye color, his eyes will be brown and the fact that he carries a gene for blue eye color will be concealed.

Such hidden variation can be revealed by breeding experimental organisms with their close relatives. When this inbreeding is done, some of the recessive alleles that have been concealed in the heterozygous state will become homozygous and will then be expressed. For example, intensive inbreeding of fruit flies has revealed they possess several recessive alleles that when the locus is homozygous result in the expression of grossly abnormal traits such as extremely short wings, deformed bristles, blindness and other serious defects.

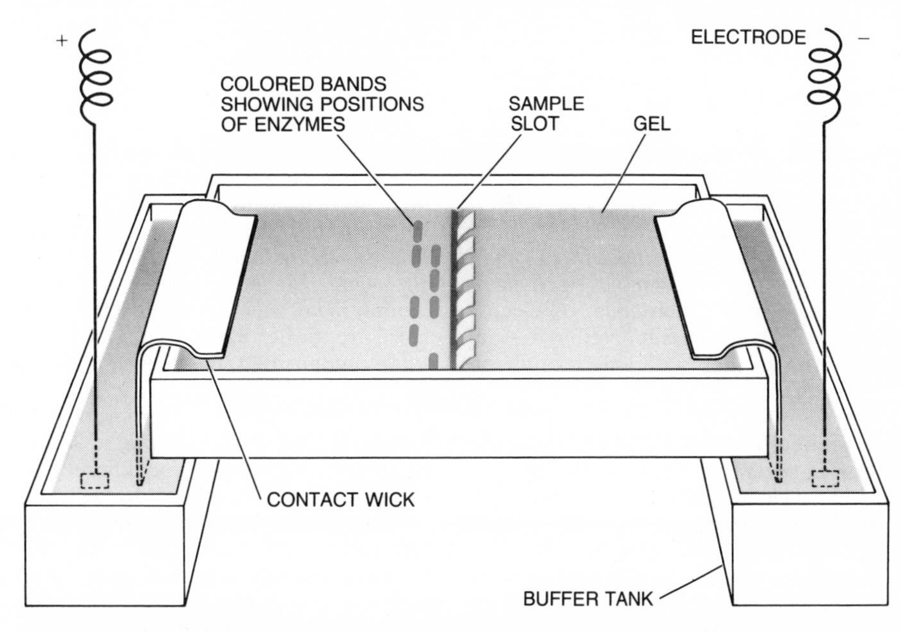

GEL ELECTROPHORESIS is a method for estimating the genetic variation of natural populations by examining the variant proteins manufactured by different individuals. First a tissue sample from each of the organisms to be surveyed is homogenized to release the proteins in the tissue; the proteins are placed on a gel of starch, agar or polyacrylamide. The gel with the tissue samples is then subjected to an electric current, usually for a few hours. Each protein in the sample migrates through the gel in a direction and at a rate that depend on its net electric charge and molecular size. After the run is over the gel is treated with a chemical solution containing a substrate that is specific for the enzyme under study and a salt. The enzyme catalyzes the conversion of the substrate into a product, which then couples with the salt to give colored bands at the points to which the enzyme had migrated. Because enzymes that are specified by different alleles may have different molecular structures and charges (and hence different mobilities in an electric field) the genetic makeup at the gene locus coding for a given enzyme can be established for each individual from the number and position of the electrophoretic bands.

Another indication of the magnitude of genetic variation in natural populations has been provided by artificial-selection experiments. In such experiments those individuals of a population that exhibit the greatest expression of a particular commercially desirable trait are chosen to breed the next generation. If a plant breeder wants to increase the yield of a variety of wheat, he will select those plants with the greatest yield at each generation and utilize their seed to grow new progeny. If the selected population changes over the generations in the direction of the applied selection, then it is clear the original plants possessed a reservoir of genetic variation with respect to the selected trait.

Indeed, the changes obtained by artificial selection are often enormous. In one flock of White Leghorn chickens egg production increased from 125.6 eggs per hen per year in 1933 to 249.6 eggs per hen per year in 1965: an increase of nearly 100 percent in 32 years! Selection can also be successfully practiced in opposite directions. For example, selection for high protein content in a variety of corn increased the protein content from 10.9 to 19.4 percent, whereas selection for low protein content reduced the protein content from 10.9 to 4.9 percent. Artificial selection has been successful in creating large numbers of commercially desirable traits in domesticated species such as cattle, swine, sheep, poultry, corn, rice and wheat, as well as in experimental animals such as fruit flies, in which artificial selection of more than 50 different traits has been accomplished. The fact that artificial selection works almost every time it is attempted indicates there is genetic variation in populations for virtually every characteristic of the organism.

This kind of evidence suggested to biologists that natural populations do have large stores of genetic variation. Yet until quite recently the limitations of traditional genetic analysis prevented investigators from determining precisely how much variation there is. Consider what would be required to find out what proportion of the genes of an individual are heterozygous. It is almost impossible to study every gene locus because of the scale of the task, but if one could obtain an unbiased sample of all the genes of an organism, it would be possible to extrapolate the values observed in that sample to the population as a whole. Indeed, opinion pollsters are able to predict with fair accuracy how millions of people will vote in a U.S. Presidential election on the basis of a representative sample of about 2,000 people: .001 percent of the population. The fact remains that with Mendelian techniques it is impossible to obtain an unbiased sample of all the genes in an individual because classical genetic analysis (involving crossbreeding of individuals exhibiting different traits) detects only those loci that are variable (that have different alleles). Since there is no way to detect invariant loci, it was impossible to obtain a truly random sample of all the genes.

The way out of this dilemma was provided by the molecular biological revolution of the past two decades. Since many genes code for proteins, one can infer variation in the genetic material from variation in the proteins manufactured by individuals. If a certain protein is invariant among the individuals of a population, the gene coding for that protein is probably also invariant; if the protein is variable, then the gene too is variable. By selecting a number of proteins that represent an unbiased sample of the genes in an organism it is therefore possible to estimate the number of alleles in a population and the frequencies at which they occur.

Biochemists have known since the early 1950's how to determine the amino acid sequence of proteins, but several months or years are usually required to sequence one protein, let alone the thousands that would be needed to obtain a statistically valid sample. Fortunately there is a simple technique, gel electrophoresis, that makes it possible to study protein variation with only a moderate investment of time and resources. Since the late 1960's this technique has been exploited to estimate the genetic variation in several natural populations.

In gel electrophoresis ground tissue or blood from several individuals is inserted into a gel consisting of starch, the synthetic polymer acrylamide or some other substance providing a homogeneous matrix. When an electric current is passed through the gel, the proteins in the tissue migrate at a rate that is determined primarily by the electric charge on their constituent amino acids (although the size and conformation of the protein may also influence the migration). Electrophoresis is so sensitive that it can detect proteins that differ by a single amino acid out of a total of some hundreds—provided that the substitution of one amino acid for another results in a change in the total electric charge on the molecule.

The proteins manufactured by different individuals in a population are compared by running them side by side in a gel for a certain time interval. The positions of the proteins after they have migrated are determined by applying a stain specific for the protein under study, which is usually an enzyme. Because each amino acid chain in a protein (some proteins have more than one chain) is the product of a single gene this approach enables the investigator to estimate how many loci in the population have multiple alleles and how many are invariant. To obtain a rough survey of variation in natural populations about 20 loci are usually examined. One useful measure of variation is heterozygosity: the average proportion of loci at which an individual in the population possesses two alleles.

Electrophoretic techniques were first applied to estimating genetic variation in natural populations in 1966, when three studies were published, one dealing with man and the other two with *Drosophila*. Since then numerous populations have been surveyed and many more are studied every year. One recent survey concerned the krill *Euphausia superba,* a shrimplike organism that thrives in the waters near Antarctica and is a major food source of whales. A total of 36 gene loci coding for different enzymes were examined in 126 krill individuals. No variation was detected at 15 loci, but at each of the other 21 loci two, three or four allelic gene products were found in the population. In other words, approximately 58 percent of the loci in this krill population had two or more alleles. On the average each krill individual was heterozygous at 5.8 percent of its loci.

Large amounts of genetic variation have been found in most natural populations studied, including 125 animal species and eight plant species. Among animals, invertebrates generally show more genetic variation than vertebrates, although there are some exceptions. The average heterozygosity for invertebrates is 13.4 percent; the average for vertebrates is 6.6 percent. The heterozygosity for man is 6.7 percent, close to the vertebrate average. Plants have a great deal of genetic variation: the average heterozygosity for eight species is 17 percent.

These estimates become even more dramatic when it is taken into account that electrophoresis underestimates genetic variation. One reason is the redundancy of the genetic code: not all mutations or substitutions in the DNA result in changes in the amino acid sequence of proteins. Moreover, since electrophoresis distinguishes proteins that have different amino acid compositions by their differential migration in an electric field, if a mutation does not alter the electrical properties of the molecule, it will not be detected. For example, if a positively charged amino acid (say glutamic acid) is replaced in a variant protein by another positively charged amino acid (say aspartic acid), the two proteins may be indistinguishable by electrophoretic criteria. Although it is clear that the estimates of variation in natural populations obtained by electrophoresis are underestimates, the degree of underestimation is not yet known. Several laboratories are now attempting to solve this problem so that genetic variation can be more precisely estimated.

In any case the extent of the variation observed in natural populations is vastly greater than that predicted by classical Darwinian theory. Instead of being homozygous for a dominant allele at most loci, individuals are heterozygous at a large proportion of loci. This fact has important consequences, particularly for animals that reproduce sexually.

Sexual reproduction involves the fusion of two germ cells (the sperm and the egg in animals), each of which possesses only one set of chromosomes instead of the two homologous sets possessed by each tissue cell. The germ cells are formed by the process of meiosis, or reduction division, in which the normal complement of chromosomes is reduced by half. In the first stage of meiosis the chromosomes duplicate themselves and the homologous chromosomes then pair up. At this stage the paired chromosomes may break in several places and exchange pieces, the process called recombination. The resulting chromosomes are a mosaic of the homologous paternal and maternal chromosomes and hence have a new combination of alleles. In the second stage of meiosis each cell divides twice to yield four germ cells. During the second division the homologous chromosomes are randomly assorted, so that there is a mixture of maternal and paternal chromosomes in each germ cell.

The scrambling of the genes by recombination (which generates new combinations of alleles on the same chromosome) and random assortment

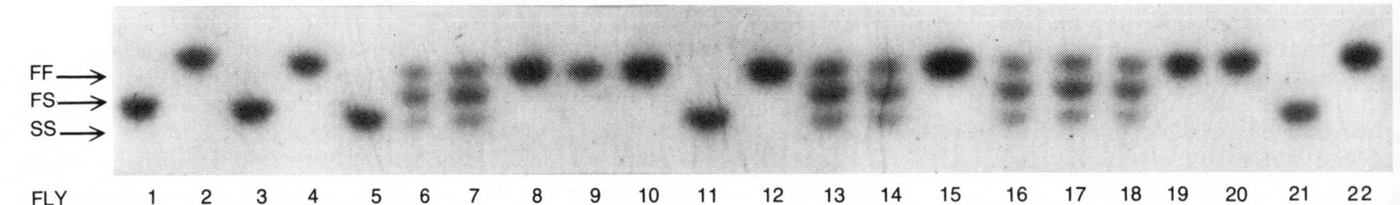

ELECTROPHORETIC GEL shown here was stained for malate dehydrogenase, an enzyme involved in the oxidation of foodstuff. The gel contains samples from 22 flies of the species *Drosophila equinoxialis*. Two variant polypeptides (the protein-chain products of two alleles) are apparent in this experiment: a fast-migrating polypeptide (designated *F*) and a slow-migrating one (designated *S*). Malate dehydrogenase consists of two polypeptide chains that combine spontaneously after they have been synthesized, so that homozygous individuals will make one form of the enzyme (either *FF* or *SS*), whereas heterozygotes will make three forms: *FF*, *SS* and *FS* (the last form has an intermediate electrophoretic mobility). Thus homozygotes exhibit only one band and heterozygotes exhibit three bands. This case, involving only two alleles in a population, is a simple one; some gene loci that code for proteins may have five alleles or more maintained in the population.

(which results in new combinations of chromosomes in the germ cells) does not in itself alter gene frequencies or cause evolution. Indeed, as it was first independently postulated by the mathematician G. H. Hardy and the biologist W. Weinberg in 1908, recombination and random assortment cause no net change in the frequencies of alleles in a population. In the absence of selection gene frequencies will remain constant from generation to generation, a hypothetical situation that has been named the Hardy-Weinberg equilibrium. The effect of recombination and random assortment is merely to reshuffle the existing genes in a population so that new combinations of alleles are exposed to selection at each generation. Sexual reproduction therefore generates a large amount of genetic diversity, greatly increasing the possibilities for evolution and providing the population with an adaptability to a changing environment far beyond the reach of an asexual species. It may be for this reason that sexuality is virtually universal in the living world, except for organisms such as bacteria, which reproduce rapidly and exist in large numbers and so may incorporate mutations in short periods of time.

Clearly the greater the heterozygosity of individuals in a sexually reproducing population is, the larger will be the number of possible combinations of alleles in the germ cells and hence in the potential progeny. Consider man, with an average heterozygosity of 6.7 percent. If we assume that there are 100,000 gene loci in man, a human individual would be heterozygous for about 6,700 loci. Such an individual could potentially produce $2^{6,700}$ ($10^{2,017}$) different germ cells, a number vastly greater than the number of atoms in the known universe (roughly estimated as being 10^{80}). Of course, such a number of germ cells will never be produced by any human individual, not even all of mankind. It follows that no two human beings ever have been or ever will be genetically identical (with the exception of identical twins and other multiple births from the same zygote, or fertilized egg). Such is the genetic basis of human individuality. The same can be said of any other organism that reproduces sexually.

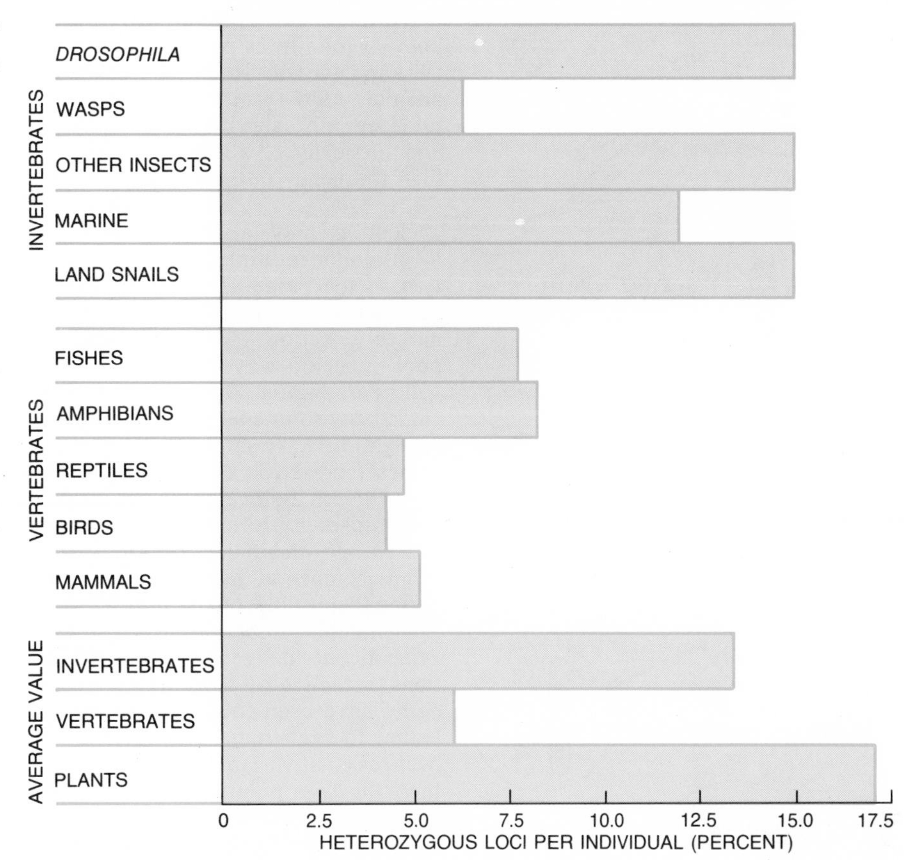

AMOUNT OF GENETIC VARIATION in natural populations, as estimated by gel electrophoresis, is surprisingly large. In general the invertebrates show more variation than the vertebrates, and the few plant species that have been studied show even more. The large numbers of alleles that are stored in the population, mostly at low frequencies, give it evolutionary flexibility.

It therefore seems clear that, contrary to Darwin's conception, most of the genetic variation in populations arises not from new mutations at each generation but from the reshuffling of previously accumulated mutations by recombination. Although mutation is the ultimate source of all genetic variation, it is a relatively rare event, providing a mere trickle of new alleles into the much larger reservoir of stored genetic variation. Indeed, recombination alone is sufficient to enable a population to expose its hidden variation for many generations without the need for new genetic input by mutation.

One can conclude that large numbers of alleles are stored in populations even though they are not maximally adaptive for that time or place; instead they are maintained at low frequency in the heterozygous state until the environment changes and they suddenly become adaptive, at which point their frequency gradually increases under the influence of natural selection until they become the dominant genetic type. But how do natural populations maintain the large reservoirs of genetic variation needed to respond to a changing environment? When one allele is locally more adaptive than another, one would expect that natural selection would gradually eliminate the less advantageous alleles in favor of the more advantageous ones until every locus is homozygous. Hence the persistence of locally disadvantageous alleles in a population can be explained only by postulating mechanisms that actively maintain diversity in spite of the selective forces tending to eliminate it.

One such mechanism is heterozygote superiority. If the heterozygote *Aa* survives or reproduces better than either homozygote *AA* or *aa*, then neither allele can eliminate the other. The most striking example of the mechanism is sickle-cell anemia. This human disease, which is prevalent in tropical Africa and the Middle East, results from an allele that gives rise to a variant form of hemoglobin, the oxygen-transporting protein in red blood cells. Biochemical studies have shown that the trait is due ultimately to the substitution of one amino acid (valine) for another (glutamic acid) at one position along two of the four constituent chains (with a total of nearly 600 amino acids) in the hemoglobin molecule; the abnormal hemoglobin can be distinguished from the normal form by electrophoresis. The slight change in the structure of the variant hemoglobin has catastrophic effects: it causes the hemoglobin molecules inside the red blood cells to form long strands. As a result the cells collapse to the shape of a sickle, resulting in a severe form of anemia that is usually fatal before reproductive age.

Since the sickle-cell allele is obviously disadvantageous, why does it persist in the human population of tropical Africa at frequencies of as high as 30 percent? It turns out that individuals who are heterozygous for the sickle-cell trait are protected against the most lethal form of malaria, whereas normal homozygotes are not. Hence the heterozygote

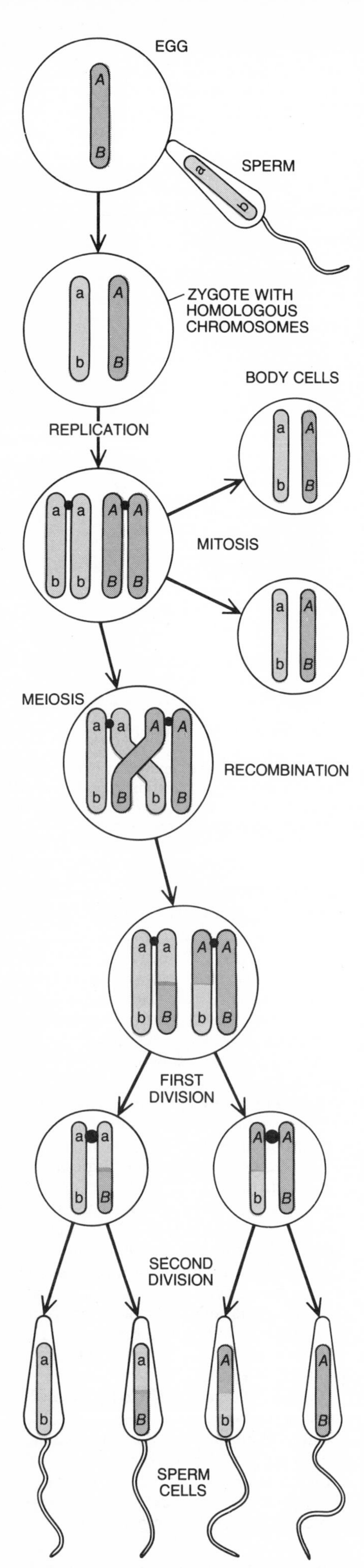

ALLELES ARE RESHUFFLED during sexual reproduction. The germ cells are formed by meiosis, or reduction division, during which the homologous chromosomes exchange corresponding segments, the process called recombination. The homologous paternal and maternal chromosomes are also randomly distributed into germ cells, so that additional combinations of alleles are created. The greater the heterozygosity of two mating individuals, the larger the number of possible sets of alleles in the germ cells and hence in the potential progeny. Meiosis does not change gene frequencies; it exposes new combinations of alleles to selection at each generation.

individual is clearly superior over either homozygote: he is protected from malaria and does not suffer from sickle-cell anemia. As a result the heterozygotes preferentially survive and reproduce and the sickle-cell allele is maintained at high frequency in the population.

Selection may also act directly to maintain multiple alleles in a population. If the range of a species encompasses several different environments, natural selection will diversify the gene pool in such a way that several alleles will be optimally adapted to the different subenvironments. Indeed, recent investigations have shown that variant enzymes (coded for by different alleles) may differ in their catalytic efficiency, in their sensitivity to temperature, acidity or alkalinity and in their response to other environmental factors, thereby rendering them subject to natural selection. For example, some variants of the enzyme alcohol dehydrogenase in populations of the fruit fly *Drosophila melanogaster* have been found to be more resistant to heat than other variants; the heat-resistant variants are commoner in the fruit-fly populations of warmer environments than they are in those of cooler environments. This finding provides strong evidence that multiple alleles may be maintained at some loci by "diversifying selection" in populations that live in heterogeneous environments. Individuals that are heterozygous at a number of loci are also usually stronger and reproductively more successful than individuals homozygous at a large number of loci; the phenomenon is known as hybrid vigor. Perhaps the manufacture of slightly variant proteins and enzymes by the heterozygote enables it to adapt to a broader range of environmental conditions or to exploit marginal habitats.

A fourth mechanism by which multiple alleles can be maintained in a population is frequency-dependent selection, in which the fitnesses of two alleles are not constant but change with their frequency. If one allele is less advantageous than the other when it is at a high frequency but gains the advantage when its frequency declines to a certain level, then the frequency of that allele will tend to stabilize at about that level.

It is also possible that some of the variation observed in proteins represents insignificant changes at the functional level that do not alter the survival or reproductive success of the organism; such mutations would then be selectively neutral. For example, although some variant enzymes (such as the variants of alcohol dehydrogenase) have been found to have different functional characteristics, others may not. If this is the case, the few variant genes that are subject to natural selection might be scattered along a chromosome, together with other variant genes that are selectively neutral. Although some of the alleles would be selected for, the majority would merely be carried along without being tested. The extent to which evolution, particularly at the molecular level, is not subject to selection is a matter of continuing debate among evolutionary biologists.

Another controversy that has been aroused by the finding of large amounts of variation in populations is the problem of genetic load. If large numbers of less fit alleles are maintained in populations by heterozygote superiority, there will be a very high probability that at each generation a zygote will be homozygous at one or more loci for a disadvantageous allele. As a result a large number of less fit zygotes might be expected, which could be a burden of mortality and infertility too great for the population to bear. Yet it must be remembered that each locus is not subject to selection separate from the others, so that thousands of selective processes would be summed as if they were individual events. The entire individual organism, not the chromosomal locus, is the unit of selection, and the alleles at different loci interact in complex ways to yield the final product. Since alleles are more likely to be tested as members of groups than as isolated units, the cost of maintaining variation in a population is actually far lower than was originally believed.

In any case there can be no doubt that the staggering amount of genetic variation in natural populations provides ample opportunities for evolution to occur. Hence it is not surprising that whenever a new environmental challenge materializes—a change of climate, the introduction of a new predator or competitor, man-made pollution—populations are usually able to adapt to it.

A dramatic recent example of such adaptation is the evolution by insect species of resistance to pesticides. The story is always the same: when a new insecticide is introduced, a relatively small amount is enough to achieve satisfactory control of the insect pest. Over a period of time, however, the concentration of the insecticide must be increased until it becomes totally inefficient or economically impractical. Insect resistance to a pesticide was first reported in 1947 for the housefly (*Musca domestica*) with respect to DDT. Since then resistance to one or more pesticides has been reported in at least 225 species of insects and other arthropods. The genetic variants required for resistance to the most diverse kinds of pesticides were apparently present in every one of the populations exposed to these man-made compounds.

The process of evolution has two dimensions: phyletic evolution and speciation. Phyletic evolution is the gradual changes that occur with time in a single lineage of descent; as a rule these changes result in greater adaptation to the environment and often reflect environmental changes. Speciation occurs when a lineage of descent splits into two or more new lineages and is the process that accounts for the great diversity of the living world.

In sexually reproducing organisms a species is a group of interbreeding natural populations that are reproductively isolated from any other such groups. The inability to interbreed is important because it establishes each species as a discrete and independent evolutionary unit; favorable alleles can be exchanged between populations of a species but cannot be passed on to individuals of other species. Since species are unable to exchange genes, they must evolve independently of one another.

The reproductive isolation of species is maintained by means of biological barriers known as reproductive isolating mechanisms. These mechanisms are of two types: prezygotic mechanisms, which impede the mating between members of different populations and so prevent the formation of hybrid offspring, and postzygotic mechanisms, which reduce the viability or fertility of hybrid offspring. Both types of isolating mechanisms serve to forestall the exchange of genes between populations.

The prezygotic reproductive isolating mechanisms are of five major types: (1) ecological isolation, where populations occupy the same territory but live in different habitats and so do not meet; (2) temporal isolation, where mating in animals and flowering in plants occur in different seasons or at different times of day; (3) ethological isolation, where sexual attraction between males and females is weak or absent; (4) mechanical isolation, where copulation in animals or pollen transfer in plants is prevented because of the different size or shape of the genitalia or the different structure of flowers, and (5) gametic isolation, where the gametes, or male and female germ cells, fail to attract each other. The spermatozoa of male animals may also be inviable in the sexual tract of females or pollen inviable in the stigma of flowers.

The postzygotic isolating mechanisms are of three major types: (1) hybrid inviability, where the hybrid zygotes fail to develop or at least to reach sexual maturity; (2) hybrid sterility, where hybrids fail to produce functional gametes, and (3) hybrid breakdown, where the offspring of hybrids have reduced viability or fertility.

All these reproductive isolating mechanisms do not act simultaneously between two species, but two or more are usually operating. Temporal isolation tends to be commoner in plants and ethological isolation commoner in animals, but even among closely related species different sets of isolating mechanisms are often operating when different pairs of species are compared. The evolutionary function of reproductive isolating mechanisms is to prevent inbreeding, but how this end is achieved depends on the opportunism of natural selection acting in the context of the specific environmental circumstances and the available genetic variation.

Clearly the waste of reproductive effort is far greater for postzygotic isolating mechanisms than it is for prezygotic ones. If a hybrid zygote is produced that is inviable, two gametes have been wasted that could have been used in nonhybrid reproduction. Even worse, if the hybrid is viable but sterile, the waste includes not only the gametes but also the resources utilized by the hybrid during its development. The waste is still greater in the case of hybrid breakdown, which involves the resources utilized by both the hybrids and their offspring. Although gametic isolation also wastes gametes, and some other prezygotic mechanisms waste energy in unsuccessful courtship or failed copulation, in general prezygotic mechanisms are much less wasteful than postzygotic ones. For this

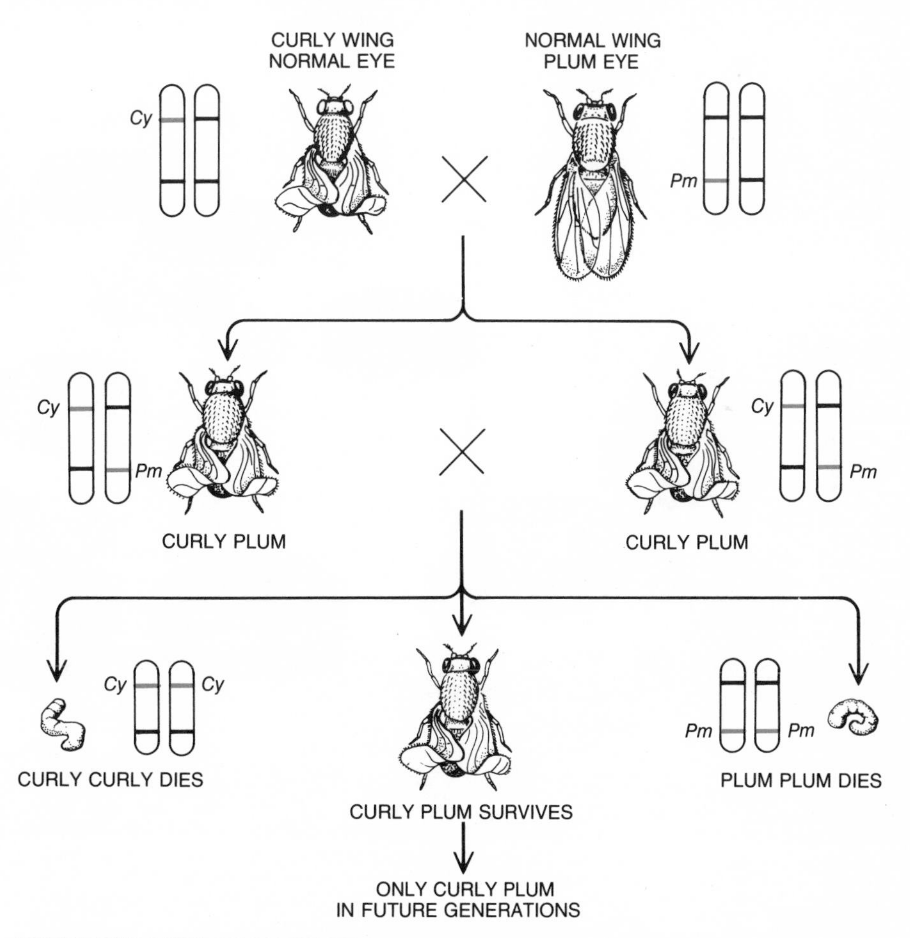

HETEROZYGOTE SUPERIORITY is one way natural selection can maintain disadvantageous alleles in a population. Shown here is a "balanced lethal" situation in *Drosophila* where homozygotes for either the "curly wing" allele or the "plum eye" allele die but the heterozygotes survive. As a result the two lethal alleles remain indefinitely in the population at frequencies of 50 percent each. A less extreme example of this mechanism in man is the case of the sickle-cell allele, which gives rise to an abnormal form of hemoglobin. Individuals who are heterozygous for the sickle-cell allele have a selective advantage over both homozygotes because they do not suffer from sickle-cell anemia (which afflicts homozygotes for the sickle-cell allele) and they are resistant to malaria (which afflicts homozygotes for the normal hemoglobin allele).

reason whenever two populations that have already been reproductively isolated by postzygotic mechanisms come in contact natural selection rapidly promotes the development of prezygotic isolating mechanisms.

Since species are reproductively isolated groups of populations, the question of how species arise is equivalent to that of how reproductive isolating mechanisms arise. Speciation commonly has two stages: a first stage in which reproductive isolation starts as an incidental by-product of the genetic divergence between two populations, and a second stage in which reproductive isolation is completed when it is directly promoted by natural selection.

The first stage of speciation requires that the exchange of genes between two populations of a species be interrupted, usually by means of a geographical separation (say the formation of a

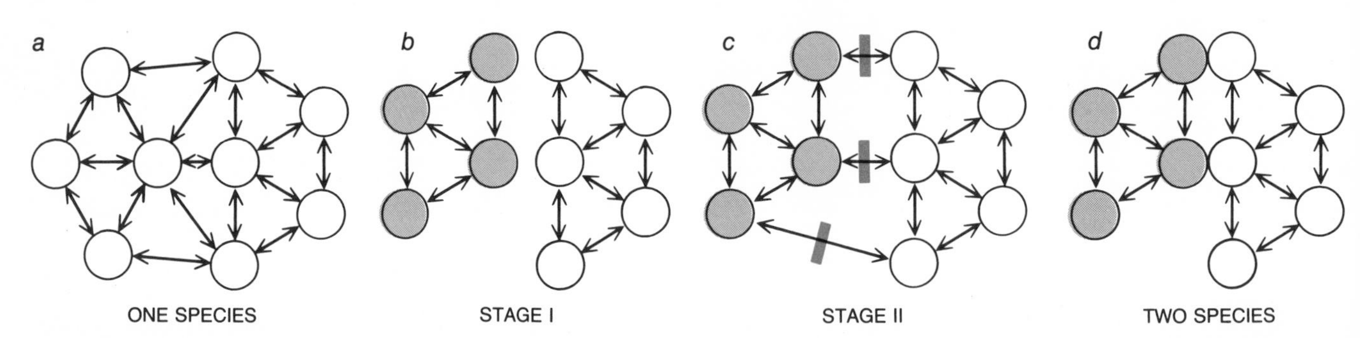

GEOGRAPHICAL SPECIATION usually occurs in two stages. In *a* local populations of a single species are represented by circles; the arrows indicate that crossbreeding may occur when individuals migrate from one population to another. Stage 1 (*b*) begins when two groups of populations become geographically isolated, so that there is no further exchange of genes between them. The isolated groups adapt to local conditions and gradually diverge genetically. In Stage 2 (*c*) individuals from the two isolated populations again come in contact. Because of the genetic divergence between the two groups crossbreeding gives rise to unviable or sterile offspring. Natural selection therefore favors the development of less wasteful prezygotic isolating mechanisms, which prevent mating between the two groups. In *d* speciation is complete: the two groups of populations coexist in the same territory without ever exchanging genes and hence evolve separately.

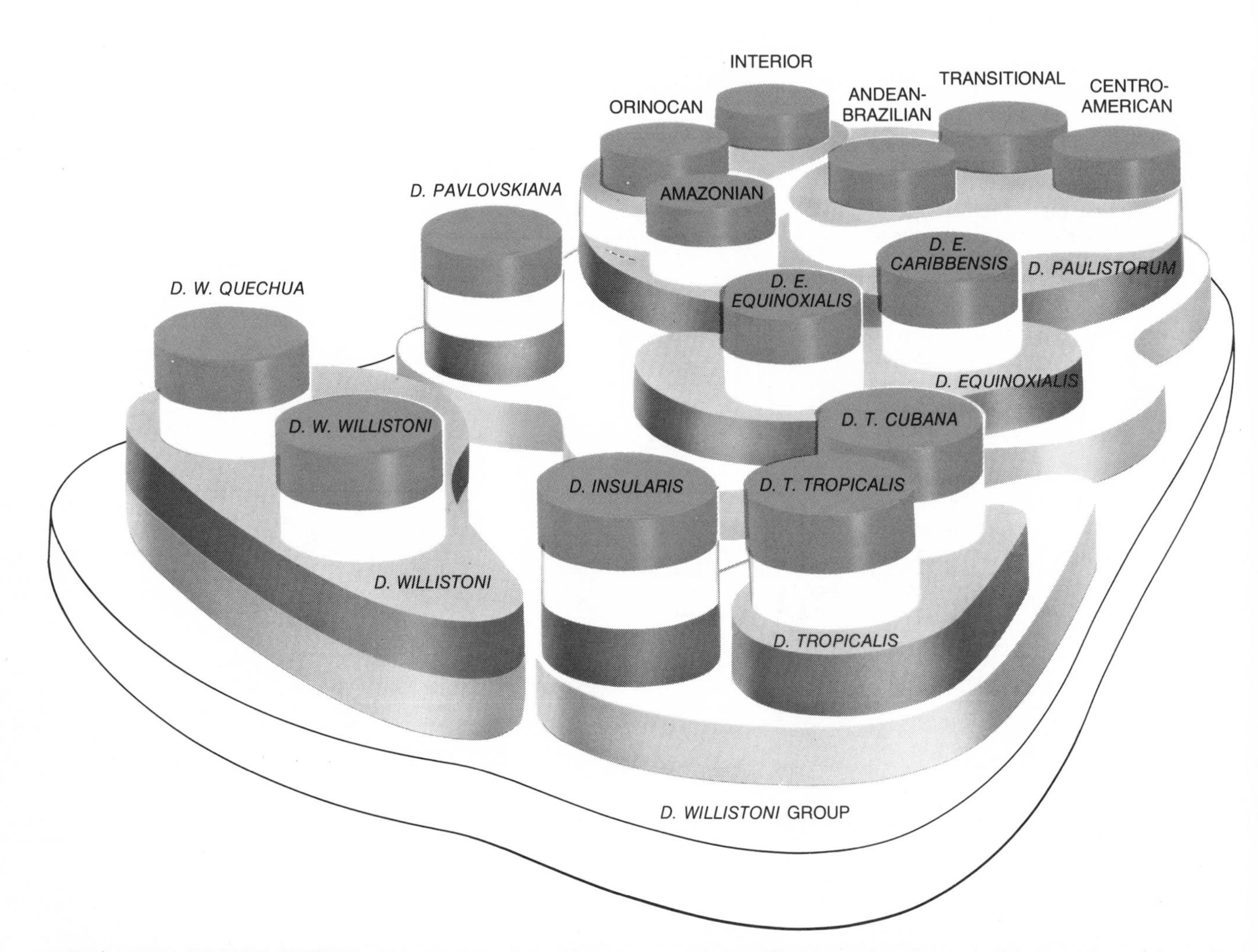

FORMATION OF NEW SPECIES within the *Drosophila willistoni* group is represented by a series of cross sections in time through the diverging phylogenetic branches. The morphologically very similar subspecies of *D. willistoni* and *D. equinoxialis* are in the first stage of speciation. The six semispecies, or incipient species, of *D. paulistorum* exhibit prezygotic isolating mechanisms and hence are in the second stage of speciation. Where two or three semispecies of *D. paulistorum* inhabit the same locality speciation is virtually complete.

mountain range between them or the emigration of one of the populations to an island). The absence of gene exchange between the two populations makes it possible for them to diverge genetically, at least in part as a consequence of their adaptation to local conditions or ways of life. As the isolated populations become increasingly different genetically, postzygotic isolating mechanisms may appear between them because hybrid offspring would have disharmonious genetic constitutions and hence a reduced viability or fertility.

The first stage of speciation is usually a gradual process, and it is often difficult to decide whether or not two populations have entered it. Moreover, the first stage may be reversible: if two populations that have been geographically isolated for some time come to have overlapping ranges, it is possible for the two populations to fuse back into a single one if the loss of fitness in the hybrids is not too great. If, on the other hand, crossbreeding yields offspring with significantly reduced viability or fertility, the populations will undergo the second stage of speciation.

The second stage involves the development of prezygotic isolating mechanisms, a process that is directly promoted by natural selection. Consider the following simplified situation. Assume that at a certain locus there are two alleles: *A*, which favors matings within the population, and *a*, which favors crossbreeding with other populations. If postzygotic isolating mechanisms operate between two populations, *A* will be common in offspring of normal fitness and *a* will be common in hybrid offspring of low fitness. As a result the *a* allele will decrease in frequency from generation to generation. Natural selection therefore favors the development of prezygotic isolating mechanisms that avoid the formation of hybrid zygotes.

Speciation can take place without the second stage if gene exchange between two populations is prevented long enough for them to diverge genetically to a significant extent. For example, the ancestors of many plants and animals now indigenous to the Hawaiian Islands arrived there from the mainland several million years ago. There they evolved and became adapted to the local conditions. Although natural selection did not directly promote reproductive isolation between the species evolving in Hawaii and the species on the mainland, the reproductive isolation of many species has nonetheless become complete.

The two stages of speciation are apparent in a group of closely related species of *Drosophila* that live in the American Tropics. The group consists of 15 species, six of which are morphologically very similar and so are termed sibling species. One of the sibling spe-

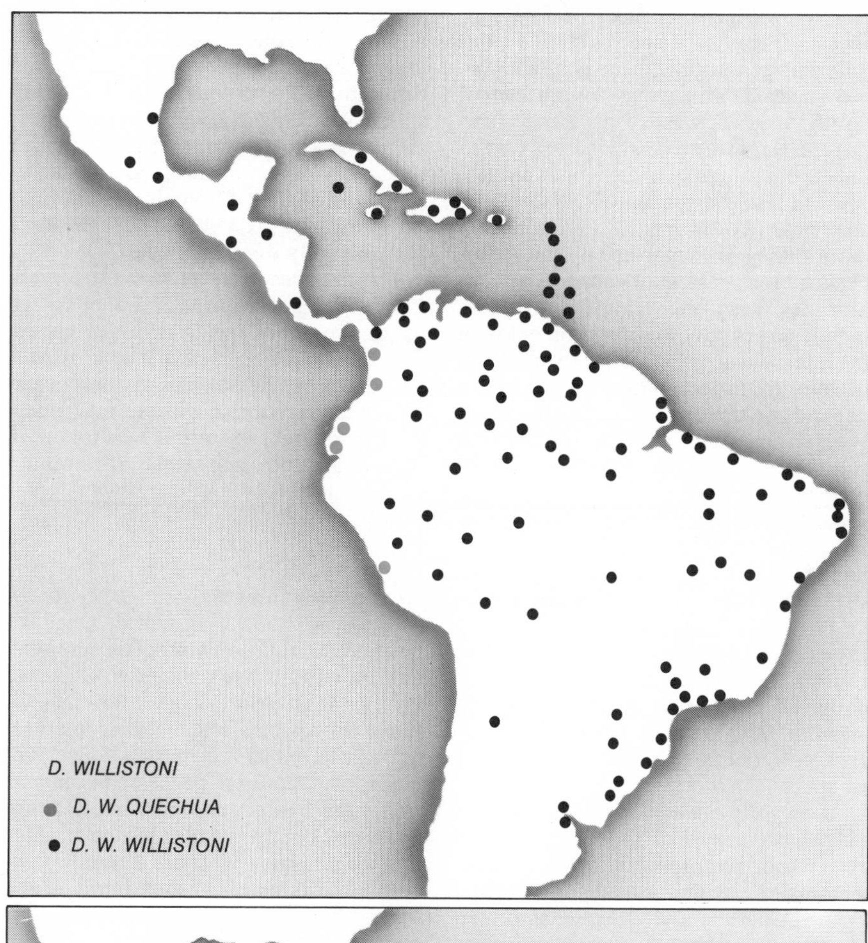

GEOGRAPHICAL ISOLATION of the subspecies of *D. willistoni* and *D. equinoxialis* is indicated on these maps. *D. willistoni willistoni* inhabits continental South America east of the Andes, Central America, Mexico and the Caribbean islands, whereas *D. willistoni quechua* inhabits South America west of the Andes. *D. equinoxialis equinoxialis* inhabits South America, whereas *D. equinoxialis caribbensis* inhabits Central America and the large Caribbean islands.

cies, *D. willistoni,* consists of two subspecies (races of a species that inhabit different geographical areas): *D. willistoni quechua,* which lives in continental South America west of the Andes, and *D. willistoni willistoni,* which lives east of the Andes and also in Central America, Mexico and the islands of the Caribbean. These two subspecies do not meet in nature; they are separated by the Andes because the flies cannot survive at high altitudes. Tests have shown that there is incipient reproductive isolation between the subspecies, particularly in the form of hybrid sterility, although the result depends on the direction of the matings. When a female *willistoni* is crossed with a male *quechua,* the male and female offspring are fertile. If, however, a male *willistoni* is crossed with a female *quechua,* the female offspring will be fertile and the males will be sterile. If these two subspecies came in geographical contact and crossbred, natural selection would favor the development of prezygotic reproductive isolating mechanisms because of the subspecies' partial hybrid sterility. The two subspecies are therefore considered to be in the first stage of speciation.

Drosophila equinoxialis is another species that consists of two geographically separated subspecies: *D. equinoxialis equinoxialis,* which inhabits continental South America, and *D. equinoxialis caribbensis,* which lives in Central America and the Caribbean islands. Laboratory crosses between the two subspecies always yield fertile female offspring and sterile male offspring, independent of the direction of the cross. Thus there is somewhat greater reproductive isolation between the two subspecies of *D. equinoxialis* than there is between the two subspecies of *D. willistoni.* Natural selection in favor of prezygotic reproductive isolating mechanisms would accordingly be stronger for *D. equinoxialis* because all the hybrid males are sterile. There is no evidence, however, of prezygotic isolating mechanisms among the subspecies of either *D. willistoni* or *D. equinoxialis,* and therefore they are not yet considered different species.

The second stage of the speciation process can also be found within the *D. willistoni* group. *Drosophila paulistorum* is a species consisting of six semispecies, or incipient species. As in *D. equinoxialis,* crosses between males and females of these semispecies yield fertile females and sterile males. In places where two or three semispecies have come in geographical contact, however, the second stage of speciation has advanced to the point where ethological isolation—the most effective prezygotic isolating mechanism in *Drosophila* and many other animals—is nearly complete. Semispecies from the same locality will not crossbreed in the laboratory but semispecies from different localities will; the reason is that the genes involved in ethological isolation have not fully spread throughout the populations. The semispecies of *D. paulistorum* therefore provide a striking example of the action of natural selection in the second stage of speciation. When ethological isolation is complete, the six semispecies will have become fully distinct species.

The final result of the process of geographical speciation can be observed in the species of the *D. willistoni* group. *D. willistoni, D. equinoxialis, D. tropicalis* and other species of this group coexist over wide territories without ever interbreeding. Hybrids are never found in nature, are extremely difficult to obtain in the laboratory and are always completely sterile.

Speciation is only one step, albeit the most fundamental one, in the diversification of the living world. Once reproductive isolation has been completed each newly formed species will take an independent evolutionary course; inevitably the species will become increasingly different as time passes. Since evolution is a gradual process, organisms that share a recent common ancestor are likely to be more similar to one another than organisms that share a remoter ancestor. This simple assumption is the logical basis of efforts to reconstruct evolutionary history by comparative studies of living organisms, which traditionally have been based on comparative morphology, embryology, cell biology, ethology, biogeography and other biological disciplines.

The task of reconstructing evolutionary history is far from simple: rates of evolutionary change may vary at different times, in different groups of organisms or with respect to different morphological features. Moreover, resemblances due to common descent must be distinguished from those due to similar ways of life, to the occupation of similar habitats or to accidental convergence. Sometimes the study of the fossil remains of extinct organisms provides clues to the evolutionary history of a group of species, but the fossil record is always incomplete and often altogether lacking.

In recent years the comparative study of nucleic acids (DNA and RNA) and proteins has become a powerful tool for the reconstruction of evolutionary history. These informational molecules retain a considerable amount of evolutionary information in their sequence of nucleotides or amino acids. Since at the molecular level evolution proceeds by the substitution of one nucleotide or amino acid for another, the number of differences in the sequence of an equivalent nucleic acid or protein in two species provides some indication of the recency of their common ancestry. One widely studied protein is cytochrome *c,* a protein involved in cell respiration; another is hemoglobin.

Investigations of evolutionary history at the molecular level have two notable advantages over comparative anatomy and other classical disciplines. One is that the information is more readily quantifiable: the number of amino acids or nucleotides that are different is readily established when the sequence of units in a protein or a nucleic acid is known for several organisms. The second advantage is that very different types of organisms can be compared. There is little that comparative anatomy can tell us about organisms as diverse as yeast, a pine tree and a fish, but there are proteins common to all three that can be compared readily.

For example, the amino acid sequence of cytochrome *c* has been determined for several organisms, from bacteria and yeast to insects and human beings. Since each amino acid substitution can involve one, two or three nucleotide substitutions in the corresponding DNA codon, one can calculate the maximum or minimum number of nucleotide changes that could have given rise to the observed amino acid substitutions. Taking the minimum number of possible nucleotide differences between the genes coding for cytochrome *c* as a basis of comparison for 20 different organisms, Walter M. Fitch and Emanuel Margoliash at Northwestern University were able to construct a phylogeny of these animals [*see illustration on opposite page*]. The overall relations agree fairly well with those inferred from the fossil record and other traditional sources. The cytochrome *c* phylogeny disagrees with the traditional one in several instances, including the following: the chicken appears to be related more closely to the penguin than to ducks and pigeons; the turtle, a reptile, appears to be related more closely to birds than to the rattlesnake, and man and monkeys diverge from the mammals before the marsupial kangaroo separates from the placental mammals.

In spite of these erroneous relations, it is remarkable that the study of a single protein yields a fairly accurate representation of the evolutionary history of 20 diverse organisms. A more accurate molecular phylogeny of these species and others should be obtained when the sequences of additional proteins and nucleic acids have been determined. The study of informational molecules from an evolutionary standpoint is a young science that was founded only about a decade ago. It is a powerful approach that should make increasingly important contributions to our understanding of biological evolution.

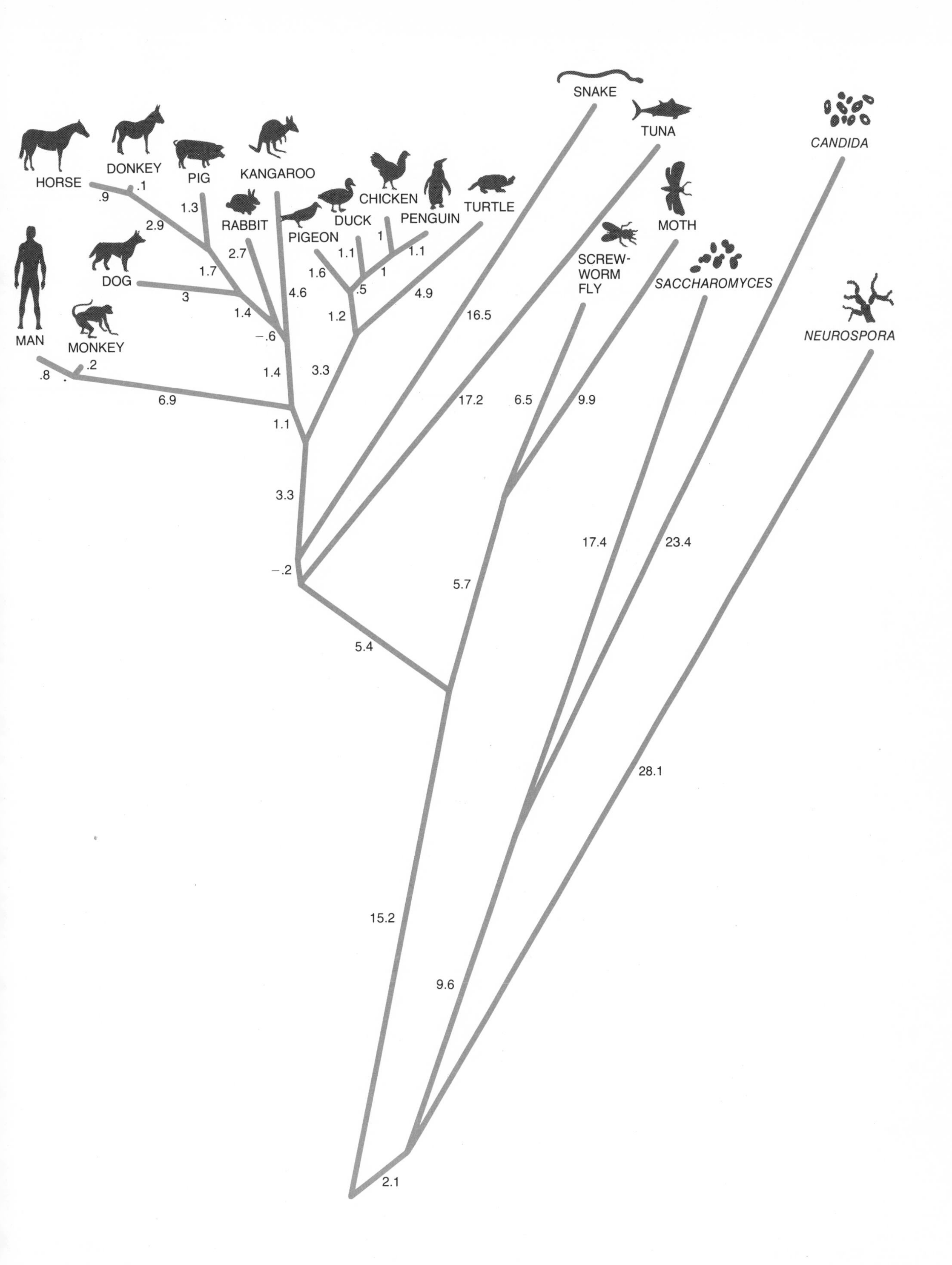

COMPUTER-GENERATED PHYLOGENY of 20 diverse organisms, based on differences in the amino acid sequence of cytochrome *c* from each species, was prepared by Walter M. Fitch and Emanuel Margoliash at Northwestern University. The phylogeny agrees fairly well with evolutionary relations inferred from the fossil record and other sources. The numbers on the branches are the minimum number of nucleotide substitutions in the DNA of the genes that could have given rise to observed differences in amino acid sequence.

III

Chemical Evolution and the Origin of Life

Chemical Evolution and the Origin of Life

BY RICHARD E. DICKERSON

Within one billion years after the formation of the earth 4.6 billion years ago one-celled organisms had evolved out of organic molecules produced nonbiologically in an atmosphere containing no free oxygen

Perhaps the most striking aspect of the evolution of life on the earth is that it happened so fast. Various radioactive-isotope methods of dating stony meteorites all give approximately the same age: 4.6 billion years. If one assumes that the sun, the planets, the meteorites and other debris of the solar system all formed from the same primordial dust cloud at about the same time, then 4.6 billion years is the age of our planet as well. Much of the earth's early geological history has been erased by later events. Some of the most ancient sedimentary rocks known are in the Fig Tree and Onverwacht deposits of South Africa, respectively 3.2 and 3.4 billion years old. Both deposits contain microfossils that resemble bacteria. Evidently some kind of primitive life had appeared on our planet a little more than a billion years after its formation.

Twice as much time was then required before the emergence of eukaryotic cells (cells with nuclei) and of multicelled organisms. The step from nonbiological organic matter to life seems to have been easier than one might have expected, and the step from one-celled bacteria to multicelled organisms seems to have been harder. The later processes are better understood because the results in the fossil record are apparent. The first billion pages in the book of the earth's history are almost completely missing. One must try to reconstruct them from information in the later pages and from what one knows about the other planets and about chemistry in general. This article describes the attempts of investigators to reconstruct the evolution of life during the missing first billion years.

Plausible mechanisms have been demonstrated for synthesizing under primitive terrestrial conditions most of the monomers, or simple molecules, needed by the living cell. Some of these monomer units are assembled into two broad classes of polymers: nucleic acids, which embody and transmit the hereditary material, and proteins, of which some serve as structural materials and others as enzymes for catalyzing the scores of complex chemical reactions that underlie both metabolism and reproduction. The problem of showing how the monomers might have linked up into biologically effective polymers has proved to be more difficult, but a number of plausible pathways have been demonstrated. Moreover, in some experiments droplets with a membrane-like boundary surface or skin have exhibited the capacity to catalyze rudimentary reactions resembling those observed in living cells, and they have demonstrated the survival advantage of being isolated from the surrounding medium. All of this is a long way from creating "life in a test tube," but that is hardly the objective. The broad goal is to arrive at an intellectually satisfying account of how living forms could have emerged step by step from inanimate matter on the primitive earth. That goal appears to be in sight.

It is possible, of course, that life did not arise on the earth at all. According to the theory of panspermia, which was popular in the 19th century, life could have been propagated from one solar system to another by the spores of microorganisms. Francis H. C. Crick and Leslie E. Orgel recently made the more venturesome suggestion that the earth, and presumably other sterile planets, might have been deliberately seeded by intelligent beings living in solar systems whose stage of evolution was some billions of years ahead of our own. The process, which Crick and Orgel call directed panspermia, might explain, for example, why molybdenum, which is quite scarce on the earth, is essential for the functioning of many key enzymes.

One can neither prove nor disprove theories of panspermia, but they are not really relevant to the inquiry of interest here. The earth is hospitable to the kind of life found on it. If that kind of life did not evolve on the earth, it must surely have evolved on a planet not drastically different from the earth in its temperature and composition. The question really is: How might life have evolved on an earthlike planet?

Assuming that terrestrial life did evolve on the earth, what was the planet like when the process began? One thing is certain: the atmosphere contained little or no free oxygen and hence was not strongly oxidizing as it is today. The organic matter that must accumulate as the raw materials from which life could evolve is not stable in an oxidizing atmosphere. One tends to forget that oxygen is a dangerously corrosive and poisonous gas, from which human beings and other organisms are protected by elaborate chemical and physical mechanisms. Many bacteria and all higher forms of life "burn" their food by combining it with oxygen, because this process yields far more energy per gram of fuel than simple anaerobic (nonoxygen) fermentation. Enzymes such as catalase, peroxidase and superoxide dismutase have evolved to protect oxygen-using organisms from toxic side effects. Anaerobic bacteria lack these protective systems; for them oxygen is both useless and lethal.

J. B. S. Haldane, the British biochemist, seems to have been the first to appreciate that a reducing atmosphere, one with no free oxygen, was a requirement for the evolution of life from nonliving organic matter. Without oxygen in the atmosphere there would have been no high-altitude ozone to block most of the ultraviolet radiation from the sun as there is today. The unblocked ultraviolet radiation reaching the surface of the planet could have then provided the energy for the synthesis of a great many organic compounds from molecules such as water, carbon dioxide and ammonia. Without free oxygen in the atmosphere to destroy them again such compounds would have accumulated in the oceans until, in Haldane's words, "the primitive oceans reached the consistency of hot dilute soup."

Haldane's ideas appeared in *Rationalist Annual* in 1929, but they elicited almost no reaction. Five years earlier the Russian biochemist A. I. Oparin had published a small monograph proposing rather similar ideas about the origin of life, to equally little effect. Orthodox biochemists were too convinced that Louis Pasteur had disproved spontaneous generation once and for all to consider the origin of life a legitimate scientific question. They failed to appreciate that Haldane and Oparin were proposing something very special: not that life evolves from nonliving matter today (the classical theory of spontaneous generation, which was untenable after Pasteur) but rather that life once evolved from nonliving matter under the conditions prevailing on the primitive earth and in the absence of competition from other living organisms.

Charles Darwin was on the right track, as he so frequently was, when he wrote to a friend in 1871: "It has often been said that all the conditions for the first production of a living organism are now present which could ever have been present. But if (and oh! what a big if!) we could conceive in some warm little pond, with all sorts of ammonia and phosphoric salts, light, heat, electricity, etc., present, that a protein compound was chemically formed ready to undergo still more complex changes, at the present day such matter would be instantly devoured or absorbed, which would not have been the case before living creatures were formed."

Harold C. Urey restated the Oparin-Haldane thesis in 1952 in his book *The Planets,* and he and Stanley L. Miller began conducting actual laboratory experiments at the University of Chicago to see whether or not energy sources available on the primitive earth could have induced the synthesis of organic compounds from gases that would have been present in the primitive atmosphere. They demonstrated that spark discharges in mixtures of hydrogen, methane, ammonia and water gave rise to aldehydes, carboxylic acids and amino acids. Other mixtures of gases, including carbon monoxide, carbon dioxide and nitrogen, were equally productive, provided that no free oxygen was present. Those experiments, and the beginnings of space exploration in the 1960's, reawakened interest in the origin of life and in the possible existence of life elsewhere in the universe.

From the outset Oparin and Haldane held divergent views about the initial conditions most important for the evolution of life, a disagreement that continues among origin-of-life theorists to-

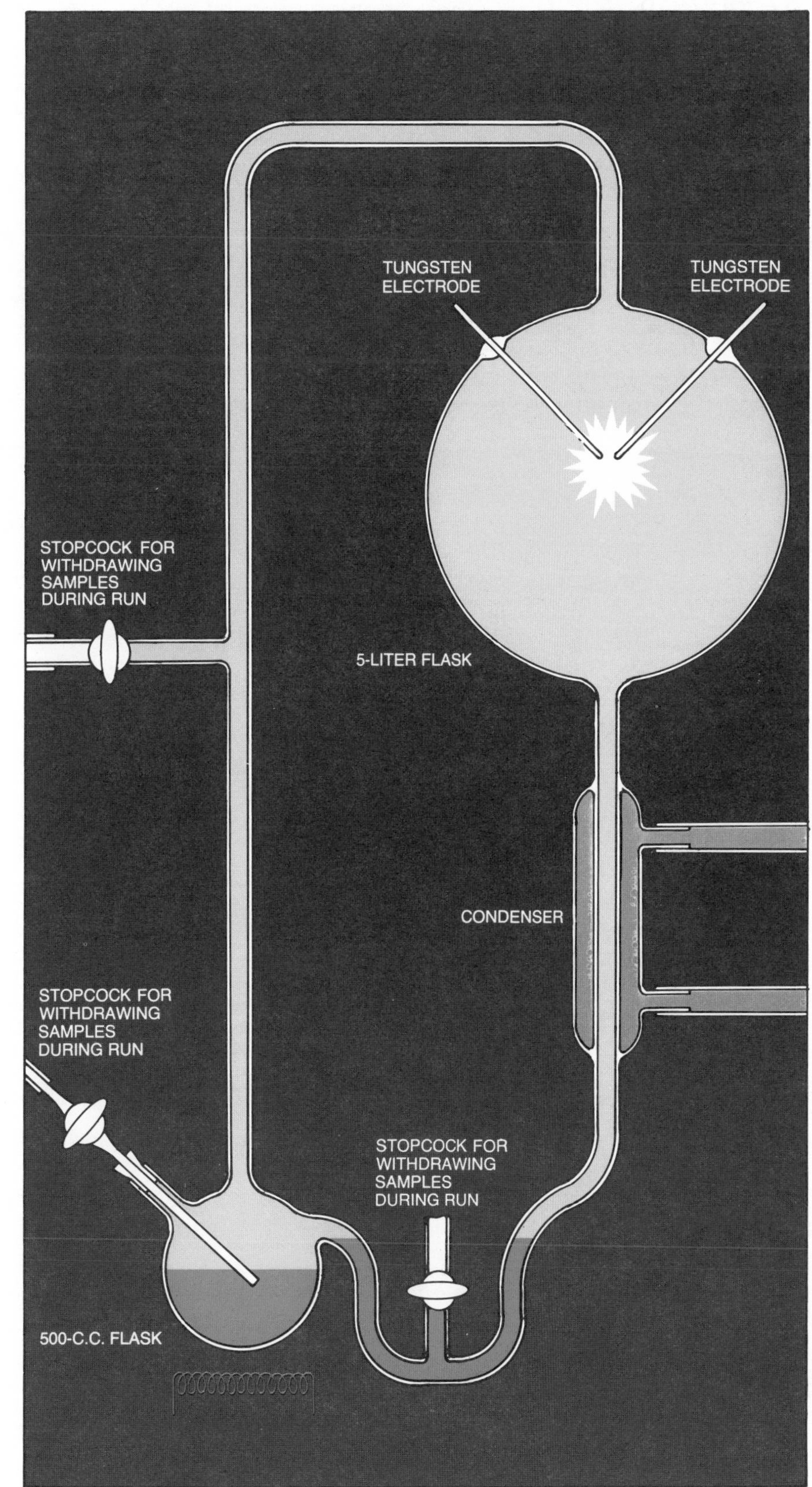

ORGANIC COMPOUNDS WERE SYNTHESIZED in an apparatus designed by Stanley L. Miller and Harold C. Urey at the University of Chicago to simulate conditions in the atmosphere of the primitive earth. Various mixtures of gases presumed to have been present in that atmosphere were admitted to the apparatus through the stopcock in the middle of the vertical tube at the left. Water in the 500-cubic-centimeter flask at the bottom of the tube was boiled to drive gases in a closed circuit through the apparatus. In the five-liter flask at the upper right the gases were subjected to a spark discharge (*white*) simulating energy inputs also presumed to have been present in the primitive atmosphere. The various compounds that were formed in the discharge (*see top illustration on page 35*) accumulated in solution at bottom of apparatus.

ELEMENT	SYMBOL	ATOMIC NUMBER	ENTIRE UNIVERSE	ENTIRE EARTH	CRUST OF EARTH	OCEAN WATER	HUMAN BODY
HYDROGEN	H	1	92,714	120	2,882	66,200	60,563
HELIUM	He	2	7,185	—	—	—	—
LITHIUM	Li	3	—	—	9	—	—
BERYLLIUM	Be	4	—	—	—	—	—
BORON	B	5	—	—	—	—	—
CARBON	C	6	8	99	55	1.4	10,680
NITROGEN	N	7	15	0.3	7	—	2,440
OXYGEN	O	8	50	48,880	60,425	33,100	25,670
FLUORINE	F	9	—	3.8	77	—	—
NEON	Ne	10	20	—	—	—	—
SODIUM	Na	11	0.1	640	2,554	290	75
MAGNESIUM	Mg	12	2.1	12,500	1,784	34	11
ALUMINUM	Al	13	0.2	1,300	6,251	—	—
SILICON	Si	14	2.3	14,000	20,475	—	—
PHOSPHORUS	P	15	—	140	79	—	130
SULFUR	S	16	0.9	1,400	33	17	130
CHLORINE	Cl	17	—	45	11	340	33
ARGON	Ar	18	0.3	—	—	—	—
POTASSIUM	K	19	—	56	1,374	6	37
CALCIUM	Ca	20	0.1	460	1,878	6	230
SCANDIUM	Sc	21	—	—	—	—	—
TITANIUM	Ti	22	—	28	191	—	—
VANADIUM	V	23	—	—	4	—	—
CHROMIUM	Cr	24	—	—	8	—	—
MANGANESE	Mn	25	—	56	37	—	—
IRON	Fe	26	1.4	18,870	1,858	—	—
COBALT	Co	27	—	—	1	—	—
NICKEL	Ni	28	0.1	1,400	3	—	—
COPPER	Cu	29	—	—	1	—	—
ZINC	Zn	30	—	—	2	—	—
			99,999.5	99,998.1	99,999	99,994.4	99,999

DISTRIBUTION OF THE MAJOR ELEMENTS varies widely according to the nature of the sample. This table shows the abundance of the first 30 elements in the periodic table in atoms per 100,000 for the entire universe, for the entire earth, for the crust of the earth, for ocean water and for the human body. The blanks indicate that the abundance is less than .1 atom per 100,000. It is apparent from the table that the earth is a highly unrepresentative sample of the elements present in the universe as a whole. The composition of the human body is fairly typical of the composition of all living organisms. Twenty-four of the elements are now known to be essential for the processes of life: the 20 given in color together with selenium (Atomic No. 34), molybdenum (No. 42), tin (No. 50) and iodine (No. 53).

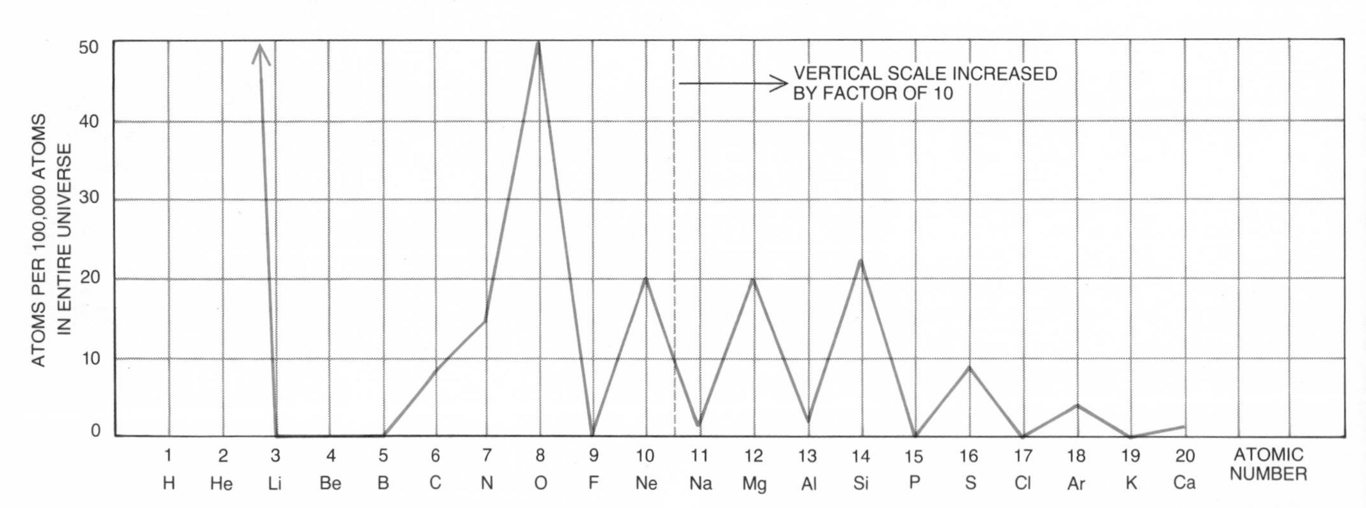

GRAPHICAL REPRESENTATION of the abundance of elements in the universe reveals the predominance of elements of even atomic number, that is, elements whose nuclei contain an even number of protons. The explanation is that the stellar nuclear reactions synthesizing the heavier elements, at least those up to iron, favor the helium nucleus as a building block. Beryllium is scarce because the fusion of three helium nuclei to form carbon is favored over the reaction that stops with the fusion of two helium nuclei to form beryllium.

day. A living cell has two central talents: a capacity for metabolism and a capacity for reproduction. The cell survives in the short run by rearranging the atoms of the compounds it ingests into molecules needed for its own maintenance. It survives vicariously over the long run by being able to reproduce itself and give rise to offspring with similar biochemical talents. Which came first, a functioning metabolism, protected by some kind of membrane against dilution and destruction by its surroundings, or a large molecule that survived by making copies of itself from materials in its surroundings? In other words, which is older, the "protobiont" or the "naked gene"? Haldane favored the latter idea. Oparin has always been more interested in the chemical reactions that can proceed inside droplets segregated from the bulk medium, and in the question of competition for survival among such droplets. (At the age of 84 Oparin perseveres in Moscow as the grand old man of origin-of-life research.) To Oparin the reproductive machinery and DNA are only the ultimate biochemical subtleties that turned metabolically competing protobionts into living cells.

The metabolism v. reproduction (or protein v. nucleic acid) argument should ultimately turn out to be as sterile as the chicken v. egg or heredity v. environment arguments of earlier generations. Today nucleic acids cannot replicate without enzymes, and enzymes cannot be made without nucleic acids. To the question, "Which came first, enzymes or nucleic acids?" the answer must be, "They developed in parallel." The catalysts needed to encourage reactions that favored the survival of particular droplets within the primitive Haldane soup, and the copying machinery to ensure that the catalysts were not lost as the droplets were broken up and dispersed by wave action and other mechanical forces, must have evolved together. The older systems did not survive because they could not compete with later improvements for their raw materials. Enzymic catalysis and DNA replication today are so thoroughly interwoven in living cells that it is hard to see what a simpler system might have been like. But as the British physicist J. D. Bernal wrote: "The picture of the solitary molecule of DNA on a primitive seashore generating the rest of life was put forward with slightly less plausibility than that of Adam and Eve in the Garden."

The step from aldehydes and amino acids nonbiologically formed to a living cell is a giant one. It is one thing to propose scenarios for the origin of life that might have been; it is another thing entirely to demonstrate that such scenarios are either possible or probable. As evidence there is a meager record of fossil microorganisms, a geological history of the planet, laboratory experiments that can demonstrate what primitive reactions might have been possible, extraterrestrial evidence for organic matter in meteorites and in spectra of interstellar dust, and the hope of detecting life that evolved independently on other planets.

We can divide the problem of the evolution of living cells from nonliving matter into five steps: (1) the formation of the planet, with gases in the atmosphere that could serve as raw materials for life; (2) the synthesis of biological monomers such as amino acids, sugars and organic bases; (3) the polymerization of such monomers into primitive protein and nucleic acid chains in an aqueous environment where depolymerization is thermodynamically favored; (4) the segregation of droplets of Haldane soup into protobionts with a chemistry and an identity of their own, and (5) the development of some type of reproductive machinery to ensure that the daughter cells have all the chemical and metabolic capabilities of the parent cells. Stated concisely, these are the problems of raw materials, monomers, polymers, isolation and reproduction.

SIX KINDS OF ATOMS in the molecules on the following pages are identified in this key. "Radical" refers to side chains that go through a series of reactions without change.

The universe as a whole consists almost entirely of hydrogen (92.8 percent) and helium (7.1 percent), with minor impurities such as nitrogen, oxygen, neon and all the other elements. Two important features should be noted in the top illustration on the opposite page: the abundance of an element decreases in general with an increase in its atomic number (equal to the number of protons in its nucleus), and atoms with even atomic numbers are more abundant than their neighbors with odd atomic numbers. The reason is that the heavier elements are synthesized from the lighter ones in the interior of stars and that this synthesis, at least of the elements with an atomic number up to that of iron, involves the capture of alpha particles, or helium nuclei, which have two protons. The even-numbered elements are more abundant because they lie in the mainstream of synthesis; the odd-numbered elements are less abundant because they are synthesized by side reactions.

It was once thought that the sun and planets of the solar system were formed by the aggregation and cooling of a cloud of hot gas. It now seems more likely that the starting point was a cloud of cold gas, dust particles and debris that became flattened by rotation and developed a protosun or concentrated core at the center. The cloud was subsequently heated by the release of gravitational energy and to a lesser extent by the natural radioactivity of some of its atoms. As the sun coalesced at the center of the revolving flat cloud other local inhomogeneities at varying distances from the center became aggregation points for the formation of the planets. The large outer planets—Jupiter, Saturn, Uranus and Neptune—probably represent a fair sample of the composition of the original cloud, since their elemental makeup is close to that of the universe at large. They are composed mainly of hydrogen, helium, methane, ammonia and water. The small inner planets—Mercury, Venus, the earth and Mars—are richer in the heavier elements and poorer in such gases as helium and neon, which could escape from the weak gravitational pull of these planets.

A combination of low gravity and high temperature led to a loss of most of the earth's volatile constituents to interplanetary space soon after the planet coalesced. Oxygen was greatly increased in relative abundance because it was locked into the nonvolatile silicate minerals; much nitrogen was lost because the nitrides are less stable and more easily converted into volatile gases. In general terms the earth consists of an iron-nickel core and a mantle that corresponds roughly in composition to the silicate mineral olivine ($FeMgSiO_4$). Only about .034 percent of the earth is carbon.

The earth became stratified into a core, a mantle and a crust as a result of the heat released as the planet was built up by accretion. The original surface may have been too hot for water to remain liquid, but as soon as the temperature had fallen below the boiling point water released from the interior by outgassing processes such as volcanism would have condensed to form the original oceans. Outgassing would have given rise to a secondary atmosphere composed of water vapor from the water of hydration of minerals, methane (CH_4), carbon dioxide (CO_2), carbon monoxide (CO) from the decomposition of metal carbides, ammonia (NH_3) and nitrogen from nitrides, and hydrogen sulfide (H_2S) from sulfides. It is from this secondary atmosphere, the character of which was reducing rather than oxidizing, that life presumably arose. As Haldane pointed out, the oxygen in the atmosphere today was put there mainly by

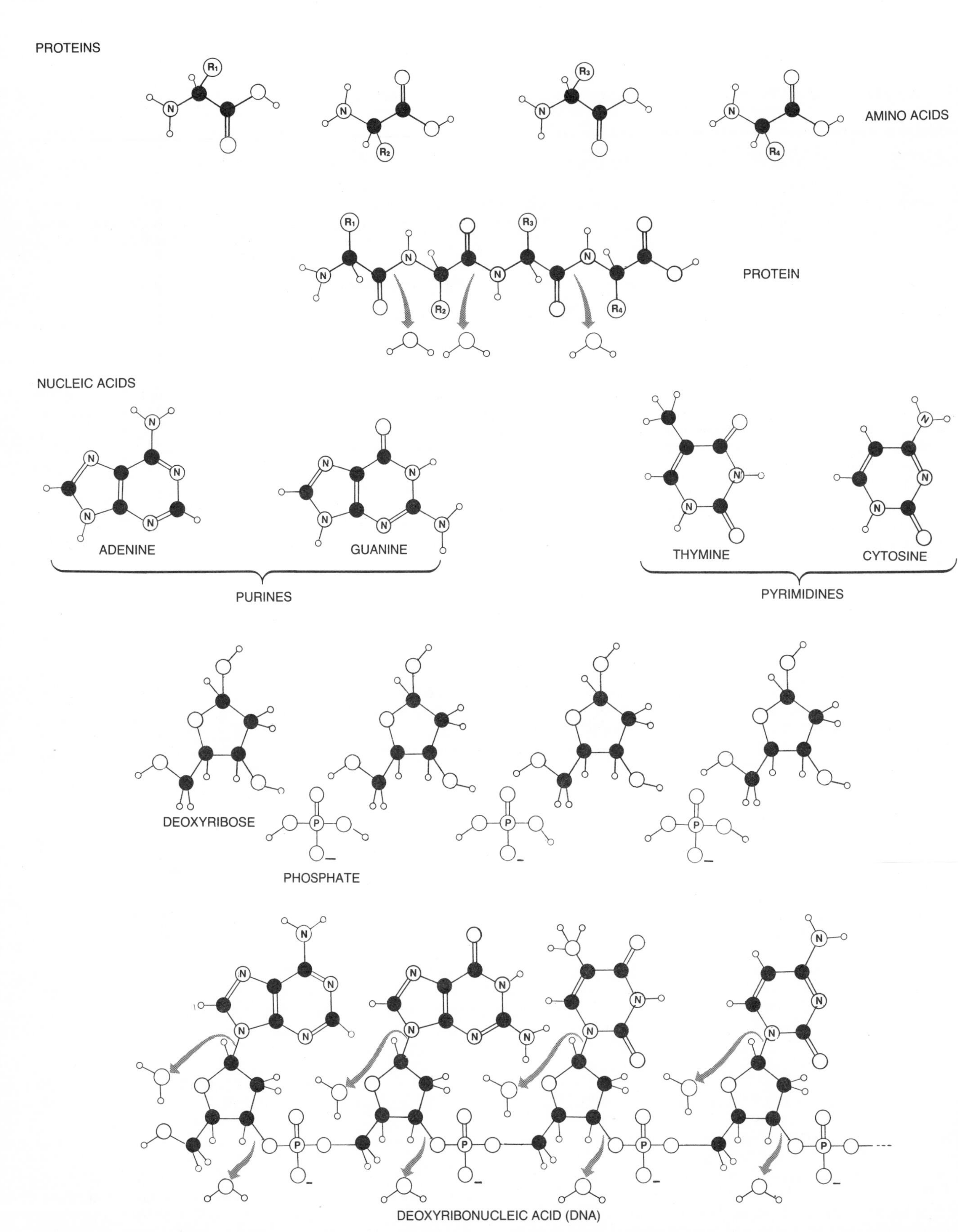

CENTRAL POLYMERS OF LIFE are proteins and nucleic acids. Proteins are polymers that serve both as structural materials and as catalysts for directing the course of biochemical reactions. Nucleic acids are polymers that embody the genetic code in which the specification for each of an organism's proteins is recorded. Deoxyribonucleic acid (DNA) has a backbone assembled from alternating units of deoxyribose (a sugar) and phosphate. Attached to each deoxyribose unit is one of four different organic bases: adenine (*A*), guanine (*G*), thymine (*T*) or cytosine (*C*). The genetic code is written in sequences of three bases (for example *ATC, GCA* or *GTA*), which represent one of 20 different amino acid monomers of a protein polymer. A sequence of triplet bases in DNA therefore specifies the sequence of amino acids in a protein. The side chains that distinguish one amino acid from another are indicated here by R_1, R_2, R_3 and so on. When amino acids are linked into polymers, a molecule of water must be removed at each linkage point, creating a peptide bond. The polymer is thus known as a polypeptide. A protein is a polypeptide of biological origin. The construction of nucleic acid chains also proceeds by removal of water molecules at the critical linkage points, a step that would require special conditions on the primitive earth.

the earliest living organisms, which succeeded in harnessing the energy of sunlight to split water molecules and fix carbon dioxide to make glucose ($C_6H_{12}O_6$), releasing oxygen as a by-product. Having once appeared on the earth, life changed the planet and destroyed the conditions that made the original appearance of life possible.

What molecules would have to be synthesized in the primitive atmosphere and oceans as the precursors of life? The list would have to include amino acids for proteins; sugars, phosphates and organic bases for nucleic acids; lipids for membranes and a number of other special-purpose organic molecules such as flavins. If polymeric chains of proteins and nucleic acids are to be forged out of their precursor monomers, a molecule of water must be removed at each link in the chain. It is therefore hard to see how polymerization could have proceeded in the aqueous environment of the primitive ocean, since the presence of water favors depolymerization rather than polymerization. We shall have to face up to this difficulty, but first let us see how the monomers could have arisen.

The formation of monomers from the gases of the primitive atmosphere is the step about which the most is known, since the reactions can be simulated and studied in the laboratory. In Miller and Urey's original experiments they worked with an artificial atmosphere consisting of hydrogen and the fully reduced forms of carbon, nitrogen and oxygen: methane, ammonia and water. Miller considered working with ultraviolet radiation in the first experiments, but difficulties with providing windows in the reaction vessel and keeping them clean of polymerized organic matter led him to conduct the first trials with an electric spark discharge, a laboratory simulation of a lightning flash.

In a typical experiment the gases were circulated past the spark for a week. The progress of synthesis was monitored by taking samples out of the boiling flask for analysis. It was surprising at first to find that the synthesized substances included several common amino acids and other molecules that are constituents of living matter. Many variations of the experiment have since been tried by Miller and by others, substituting carbon monoxide or carbon dioxide for methane, nitrogen for ammonia and ultraviolet radiation for the spark discharge. Many naturally occurring amino acids have been found, including leucine, isoleucine, serine, threonine, asparagine, lysine, phenylalanine and tyrosine. It is clear that amino acids would have been readily synthesized in the primitive atmosphere.

Two cautionary comments are necessary. Although the simulations yield

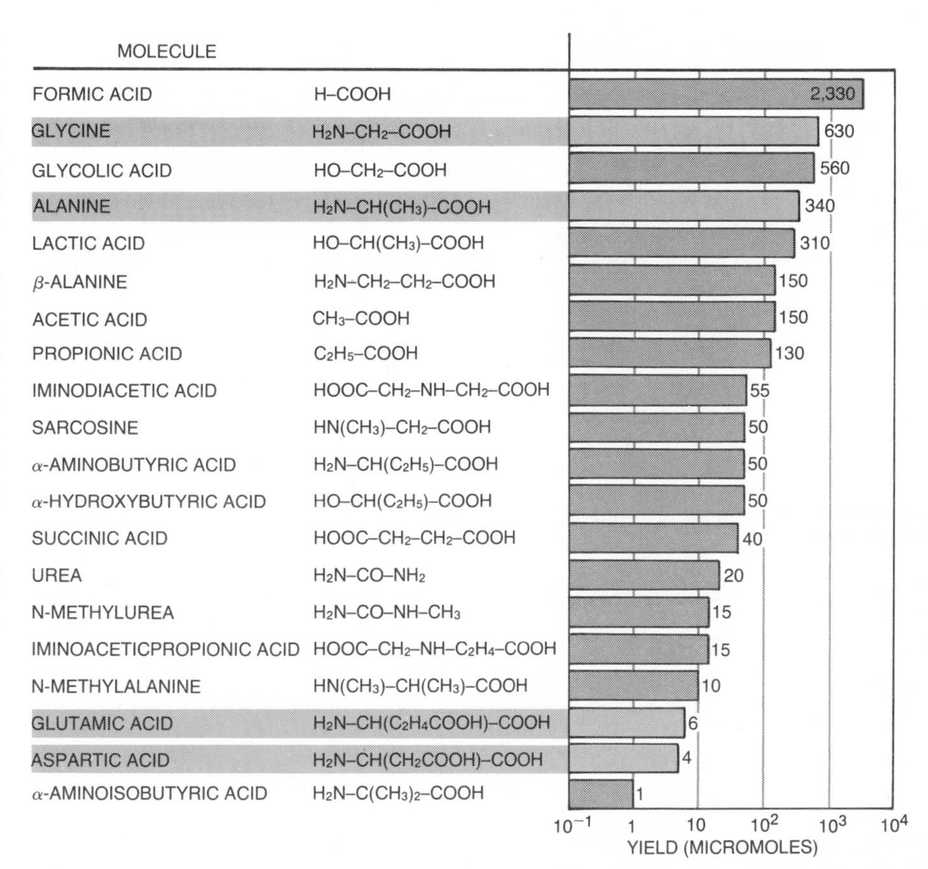

MOLECULES ASSOCIATED WITH LIFE were produced in the experiments conducted by Miller and Urey. The compounds listed here were created in one such experiment by repeatedly passing the spark discharge through a gaseous mixture of hydrogen, methane, ammonia and water. The mixture originally held 710 milligrams, or 59,000 micromoles, of carbon in the form of methane gas, of which about 15 percent was converted into the compounds listed. A substantially larger percentage of the carbon was deposited as a tarry residue that could not be analyzed. This particular experiment yielded four of the 20 amino acids that are commonly present in proteins, identified here in color. Structures of these compounds appear on page 41.

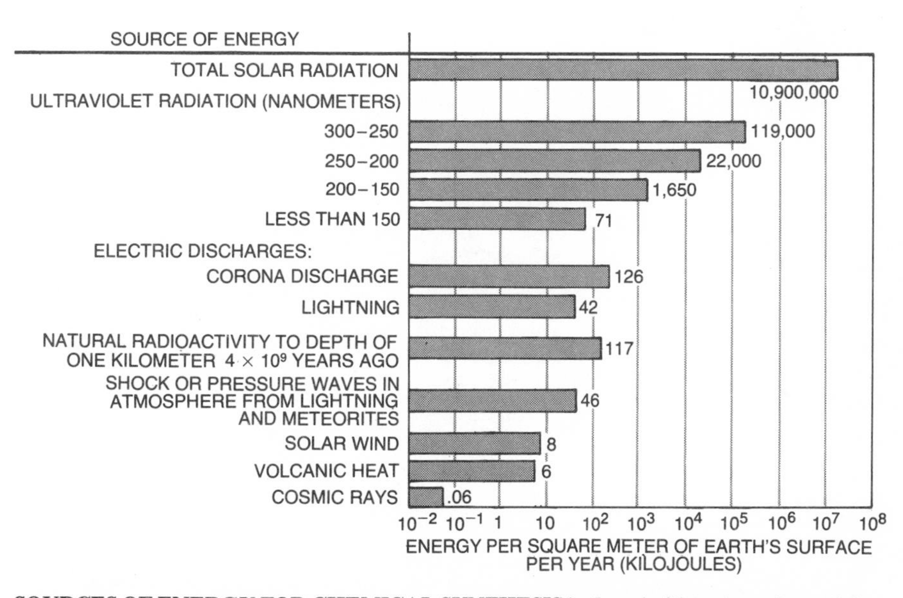

SOURCES OF ENERGY FOR CHEMICAL SYNTHESIS in the primitive atmosphere of the earth embrace a wide variety of phenomena. More than 98 percent of the energy in solar radiation arrives in the form of photons too weak to make and break chemical bonds. Only ultraviolet photons with wavelengths shorter than about 200 nanometers (less than 1.2 percent of the total ultraviolet) would have been highly effective in triggering chemical reactions. Value for energy in electric discharges, a total of 168 kilojoules per square meter of surface per year, is based on present-day weather conditions and could have been larger on the primitive earth.

many of the amino acids found in the proteins of living organisms, they also yield at least as many related molecules that are not present. For example, experiments of the Miller type synthesize three isomeric forms of an amino acid with the formula $C_3H_7NO_2$: alanine, beta-alanine and sarcosine. Yet only alanine has been incorporated into the proteins of living organisms. Of the three isomers valine, isovaline and norvaline only valine appears in proteins today. Seven amino acid isomers with the formula $C_4H_9NO_2$ are created in spark-discharge experiments, none of which is designated as a protein constituent by the universal genetic code of terrestrial life. It is obvious that the choice of the 20 amino acids in the genetic code was not foreordained by the availability of a particular set of molecules on the primitive earth. One of the fascinating side issues of origin-of-life biochemistry is why the present set of 20 amino acids was chosen. Were there false starts, with genetic codes that specified different sets of amino acids, in lines of development that died out without a trace because they could not compete with the lines that survived? There probably were.

The other cautionary observation is that these laboratory simulations of prebiological reactions give rise to equal numbers of both forms of optically active molecules: molecules that rotate polarized light in opposite directions because the molecules exist in two configurations that are mirror images of each other. Such molecules are designated by the prefix D or L, abbreviations for dextro and levo, designating the direction of rotation of the polarized light. Except for certain special adaptations involving bacterial cell walls and biochemical defense mechanisms, all living organisms today incorporate only L amino acids. Various attempts have been made to explain why only one optical isomer of amino acids came to be favored, ranging from the asymmetric crystal structure of minerals that could act as surface catalysts to the natural polarization of cosmic rays and Coriolis forces arising from the rotation of the earth (which differ in the two hemispheres). It seems likely that the primitive selection of the L isomers over the D isomers was a matter of chance. We do know that enzymes must bind molecules to their surface and that enzymes will be more efficient if they are designed to bind only one isomer or the other. There may at one time have been primitive life or precursors of life based on both D- and L-amino acids, with a 50 percent chance that the L amino acids would eventually prevail.

What are the detailed chemical steps by which amino acids are synthesized by a spark discharge or ultraviolet radiation? In following the appearance and disappearance of intermediates during week-long synthesis runs Miller and Urey observed that the concentration of ammonia fell steadily and that its nitrogen atoms appeared first in molecules of hydrogen cyanide (HCN) and cyanogen (C_2N_2), which along with aldehydes were the first substances formed. Amino acids were synthesized more slowly at the expense of hydrogen cyanide and aldehydes. This progression suggests that the amino acids are formed from aldehydes by a mechanism well known to organic chemists, the Strecker synthesis [*see equations 1, 2 and 3 in illustration on this page*].

The aldehyde first adds ammonia and loses water to form an imine; the imine then adds hydrogen cyanide to form an aminonitrile. These two steps are freely reversible. The amino acid is formed by the irreversible hydrolysis of the aminonitrile, with the addition of two molecules of water and the loss of ammonia. On the primitive earth the aminonitriles could have been synthesized in the atmosphere and dissolved and hydrolyzed in the ocean. In laboratory applications of the Strecker synthesis hydrolysis is carried out in solutions that are either acidic or basic because the rate of hydrolysis in a neutral solution is low. On the primitive earth, however, the hydrolysis could have been stretched over tens of thousands of years without penalty since there would have been no free oxygen to degrade the aminonitriles. Hydroxyacids also are formed by the Strecker synthesis. Formaldehyde (CH_2O) is converted into glycolic acid ($C_2H_4O_3$) and the amino acid glycine ($C_2H_5NO_2$), and acetaldehyde (C_2H_4O) is converted into lactic acid ($C_3H_6O_3$) and the amino acid alanine ($C_3H_7NO_2$). Some of the more complex amino acids require more complex aldehydes as starting materials. On the other hand, serine ($C_3H_7NO_3$), an amino acid with a hydroxyl group (OH), can be made by the condensation of two molecules of formaldehyde followed by a Strecker synthesis [*see equations 4 and 5 in top illustration on opposite page*]. Other special pathways have been proposed for the synthesis of most of the naturally occurring amino acids.

Electric discharges and ultraviolet radiation are not the only conceivable sources of energy for prebiological synthesis. Other sources include the emanations from radioactive elements in surface rocks and shock waves from light-

1
ALDEHYDE + HYDROGEN CYANIDE + WATER → AMINO ACID

IN ATMOSPHERE
2
+NH3
−H2O
+HCN
ALDEHYDE
IMINE
AMINONITRILE

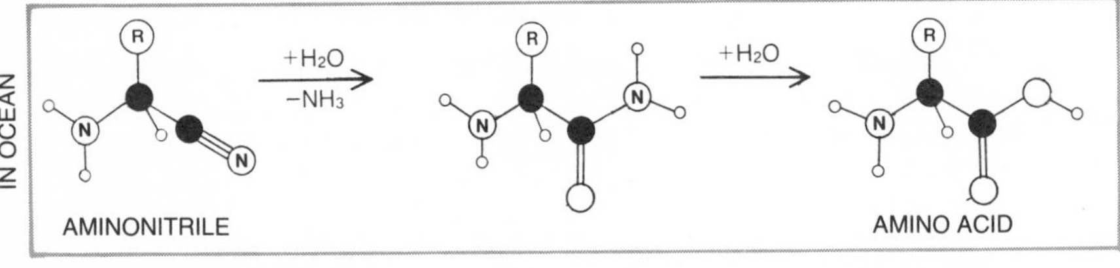

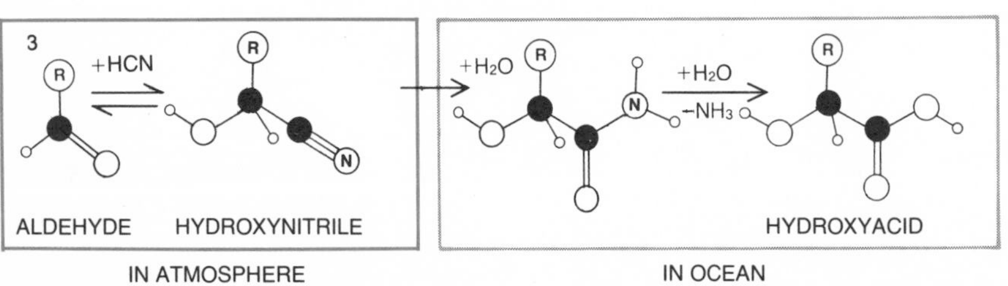

SOURCES OF AMINO ACIDS in experiments simulating the primitive atmosphere are aldehydes, hydrogen cyanide (HCN) and water (*1*). In the reaction *R*, a side chain of the aldehyde, appears as the side group of the amino acid. The reaction probably follows the steps of what is known as the Strecker synthesis (*2*). The first three steps (*color*), in which water is removed, could have taken place in the primitive atmosphere. If the aminonitrile formed entered ocean, the final hydrolysis steps could proceed. Hydroxyacids, such as lactic acid and glycolic acid, which have an additional hydroxyl group (OH), can also be formed by a Strecker synthesis (*3*).

ning and meteors. Although by far the greatest amount of energy comes from the sun, much of it is in the visible and infrared regions of the spectrum, where the photons do not have enough energy to make and break chemical bonds. Moreover, most of the ultraviolet radiation would have been ineffective in triggering chemical synthesis because methane and other small hydrocarbon molecules, water, carbon monoxide and carbon dioxide can absorb only wavelengths shorter than about 200 nanometers, or less than 1.2 percent of the available ultraviolet radiation (about 1,720 kilojoules out of a total of 143,000 kilojoules per square meter of the earth's surface per year). Of the gases presumed to have been present in the primitive atmosphere only ammonia and hydrogen sulfide can absorb longer wavelengths: ammonia up to 220 nanometers and hydrogen sulfide up to 240. As a result the two gases may have been important in the primitive atmosphere as collectors of solar energy.

If the weather on the primitive earth was no more violent than today's, lightning and the corona discharges of atmospheric electricity would have provided about 170 kilojoules per square meter per year. The amount of energy from natural radioactivity, calculated by extrapolating backward to the early years of the earth's history, would have been about 117 kilojoules. Shock waves in the atmosphere would have provided on the order of 46 kilojoules, again assuming that the primitive weather resembled our own. The "wind" of energetic particles from the sun and volcanism together may have contributed another 14 kilojoules, perhaps more if the primitive earth were more active tectonically than it is today. Of all the energy sources for prebiological synthesis electric discharges probably were the most significant, both because of the amount of energy involved and because this energy would be released close to the surface of the ocean where the products could easily be dissolved in the water.

Twenty amino acids are enough building blocks for proteins. For nucleic acids one must have two kinds of sugar (ribose for RNA and deoxyribose for DNA), phosphate and two kinds of nitrogen-containing bases: purines and pyrimidines. The sugars can be built up as condensation products of formaldehyde by the synthesis known as the formose reaction. The mechanism involves several steps, but the overall reaction is simple: five molecules of formaldehyde combine to form one molecule of ribose [*see equation 6 in middle illustration at right*]. There are problems with the formose reaction (the ribose produced is unstable in aqueous solution and the experimental conditions are not realistic simulations of primitive earth conditions), but some such reaction could

SYNTHESIS OF AMINO ACID SERINE in primitive atmosphere could have begun with condensation of two molecules of formaldehyde to form glycoaldehyde (*4*). Strecker synthesis could have converted glycoaldehyde into serine, which has hydroxyl group in its side chain (*5*).

SYNTHESIS OF SUGAR RIBOSE, a constituent of nucleic acids, can be achieved by a multistep reaction in which five molecules of formaldehyde combine to form one molecule of ribose.

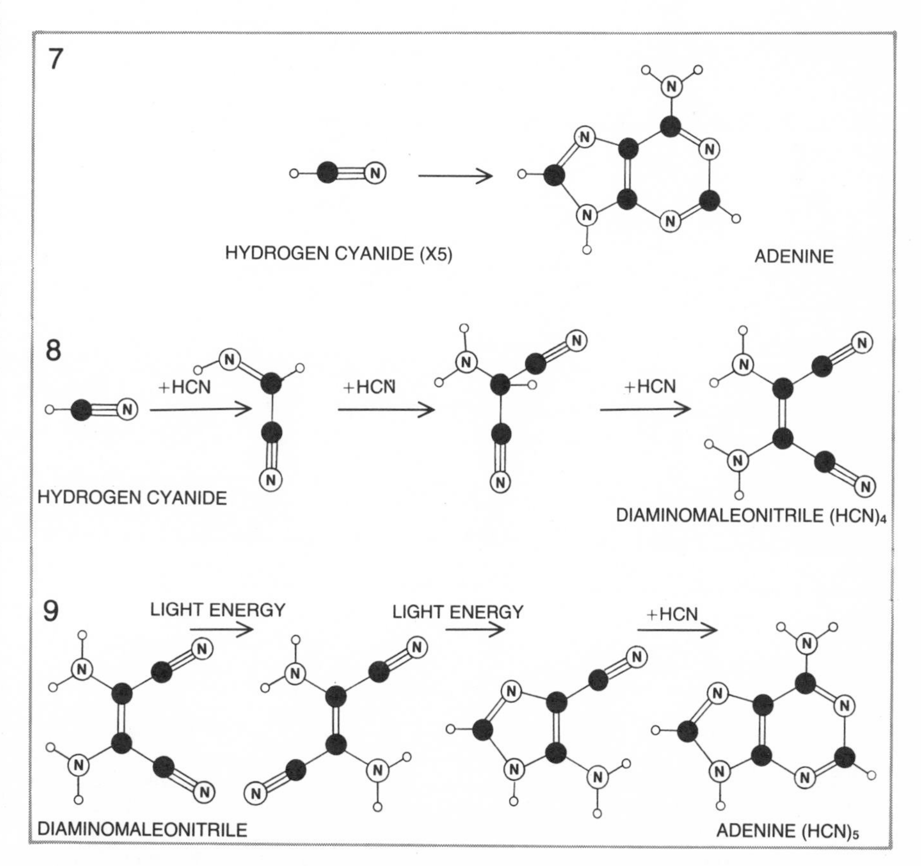

ADENINE, a pentamer of hydrogen cyanide (*7*), is the most easily synthesized of the four organic bases that serve as the coding units of DNA. Presumably four molecules of hydrogen cyanide combine to form a tetramer, diaminomaleonitrile (*8*). The tetramer rearranges itself so that it forms a five-member ring. A fifth molecule of hydrogen cyanide closes a second ring (*9*).

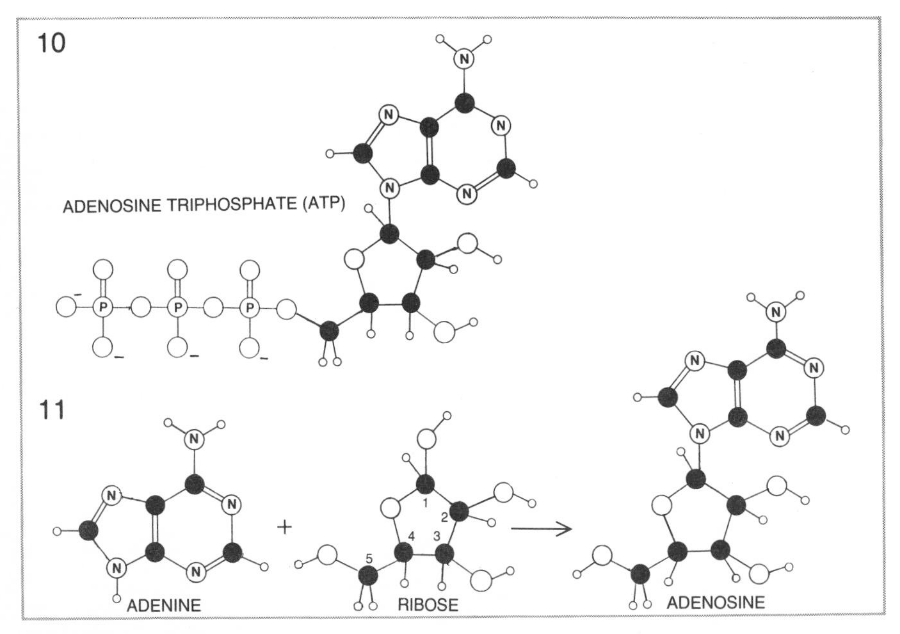

ADENOSINE TRIPHOSPHATE (ATP), the principal medium for the storage and exchange of energy in all living organisms, is created from adenine, ribose and a triphosphate tail (*10*). The nonbiological synthesis of the adenosine presents a special difficulty because the adenine might be coupled to any one of the four carbons in the ribose (1′, 2′, 3′ or 5′) that carries a hydroxyl group (*11*). Moreover, three of the four hydroxyl carbons (1′, 2′ and 3′) are asymmetric, so that alpha and beta forms of the molecule can be synthesized at each of them. In organisms today adenine and ribose are coupled at the 1′ carbon in the molecule's beta configuration.

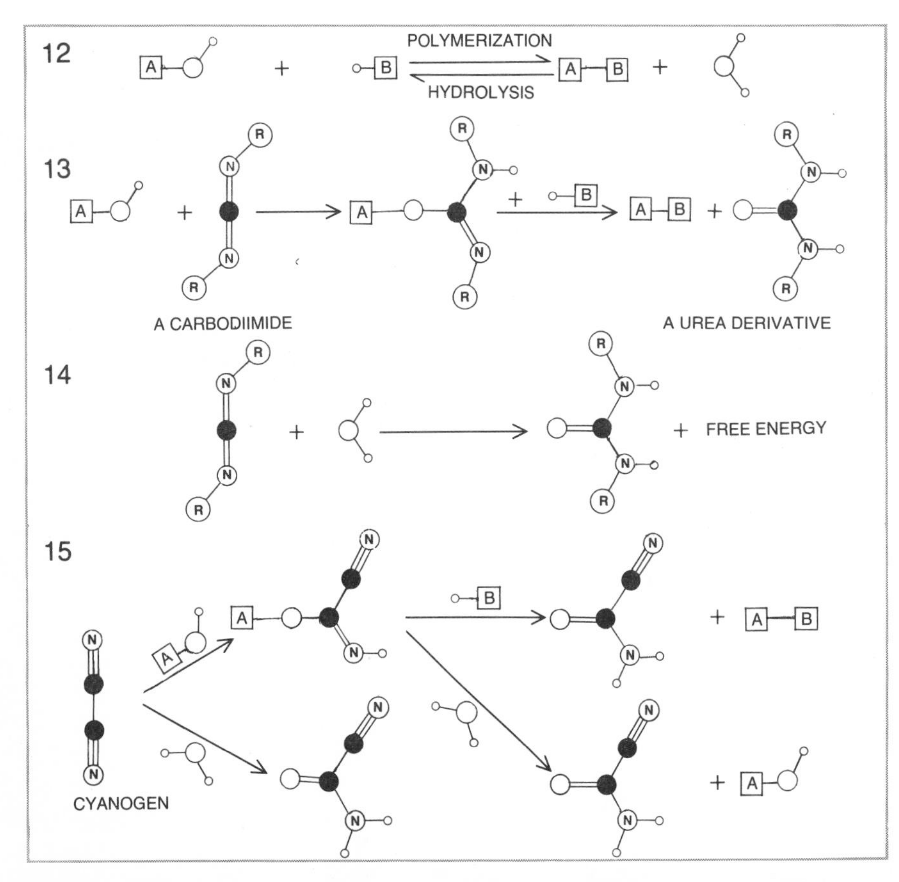

FORMATION OR LENGTHENING OF A POLYMER CHAIN requires the removal of the constituents of water from the ends of the subunits being joined. Here the subunits, which can be either monomers or polymers, are labeled *A* and *B* (*12*). In the presence of water the reaction is driven to the left, thereby uncoupling the subunits. To drive the reaction to the right requires not only the removal of water but also an input of energy. On the prebiological earth a coupling agent such as a carbodiimide could have brought about such polymerizations (*13*). The uptake of water by the carbodiimide releases the free energy needed for the reaction (*14*). Polymerization can be thwarted, however, if water is present (*15*). In this example the coupling agent is cyanogen. At each step the reactions with water are thermodynamically favored.

have yielded the necessary ribose molecules.

Among the organic bases the purine molecule adenine is the most readily synthesized. It is simply a pentamer of hydrogen cyanide: 5HCN yields $C_5H_5N_5$. It seems likely that four molecules of hydrogen cyanide combine initially to form a tetramer of HCN, diaminomaleonitrile. The diaminomaleonitrile molecule is an important intermediate in many reactions leading to base synthesis. In the presence of light the molecule can rearrange itself and add one more hydrogen cyanide to form adenine [*see equations 7, 8 and 9 in bottom illustration on preceding page*]. The synthesis proceeds under conditions that would have been reasonable on the primitive earth. Guanine, the other purine needed in nucleic acids, can be obtained from diaminomaleonitrile by hydrolysis reactions involving cyanogen. Other less convincing syntheses have been proposed for the pyrimidine bases: thymine, uracil and cytosine.

When adenine is joined to a molecule of ribose, the product is adenosine, a nucleoside. With the simple addition of a triphosphate tail adenosine becomes adenosine triphosphate (ATP), the molecule that serves as the primary currency of energy exchange in all living organisms [*see equation 10 in top illustration at left*]. It is noteworthy that the nucleoside selected for coupling to the triphosphate is adenosine and not guanosine, cytidine or uridine. There is no obvious reason why ATP is better suited for energy storage than GTP, CTP or UTP. It may be that the relative simplicity of the synthesis of adenine led to its being present in greater concentrations than the other bases in the primitive Haldane soup. The use of ATP may represent nothing more than another cosmic throw of the dice.

It is not difficult to account for the appearance of the bases and sugars of nucleic acids on the primitive earth. An unexpected stumbling block arises, however, when one tries to account for the particular way in which the bases and sugars are joined to make nucleosides, such as the coupling of adenine and ribose to form the adenosine molecule [*see equation 11 in top illustration at left*].

The problem is that ribose has four hydroxyl groups, any one of which could serve as the linkage point to adenine. Moreover, three of the hydroxyl groups are joined to "asymmetric" carbon atoms, so that adenine could be linked to each of the carbon atoms (designated 1′, 2′ and 3′) in two structurally different ways, yielding either an alpha or a beta configuration. No one has yet proposed a convincing method of getting good yields of the beta-1′ connection between adenine and ribose that is universally found in DNA and RNA.

In spite of all these qualifying state-

16

ADENOSINE TRIPHOSPHATE (ATP)

WATER

ADENOSINE DIPHOSPHATE (ADP)

PHOSPHATE

17

POLYPHOSPHATE

ATP IS AN EFFECTIVE STOREHOUSE of free energy because a large amount of energy is released when it is hydrolyzed by water to adenosine diphosphate (ADP) and a phosphate ion (*16*). This energy arises in part because the repulsion between the three bound phosphates in ATP can be relieved by breaking the connecting bonds and letting the phosphates move apart. For the same reason polyphosphate chains can also store energy (*17*). Early organisms might have relied on nonbiologically formed polyphosphates for their energy.

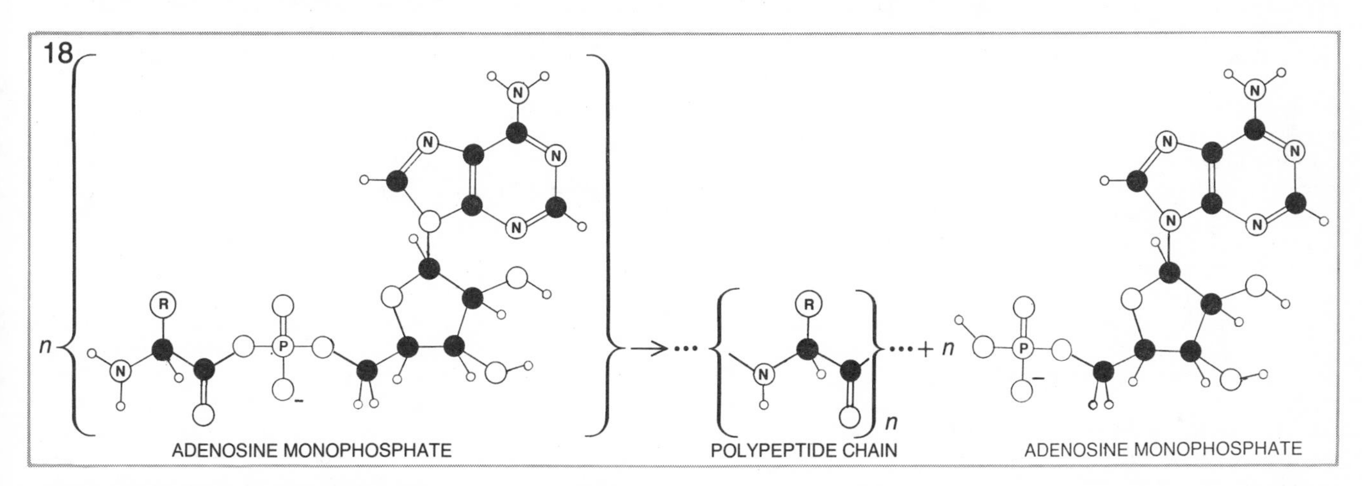

FORMATION OF POLYPEPTIDE CHAINS has been shown to take place on the surface of particles of clay. In a typical example montmorillonite clay promotes the polymerization of polypeptides from amino acid adenylates, esters formed by the reaction of amino acids with adenosine monophosphate (AMP). Energy released by phosphate ions enables the polymer to form in the presence of water.

19 GREEN AND PURPLE SULFUR BACTERIA

LIGHT ENERGY

HYDROGEN SULFIDE (X12) + CARBON DIOXIDE (X6) → + SULFUR (X12) + WATER (X6)

20 CYANOBACTERIA, ALGAE, PLANTS

LIGHT ENERGY

WATER (X12) + CARBON DIOXIDE (X6) → GLUCOSE + OXYGEN (X6) + WATER (X6)

INVENTION OF PHOTOSYNTHESIS enabled living organisms to become the primary producers of energy-rich molecules instead of mere consumers of those provided by prebiological processes. The first organisms capable of photosynthesis probably made use of hydrogen sulfide (H_2S) as a source of hydrogen atoms for converting carbon dioxide (CO_2) into glucose (*19*). Later the predecessors of the cyanobacteria (blue-green algae), green algae and higher plants mastered the technique of obtaining hydrogen from water, a more elaborate two-step process (*20*) that released oxygen and transformed the atmosphere into the one in which all subsequent life developed.

ments, current knowledge of the chemistry by which amino acids, bases, sugars and other monomers of life could have been synthesized on the primitive earth is really rather impressive. Although the answers are not yet known, the problems can at least be defined, and they seem to be chemical problems that are likely to yield to further effort and experimentation. There are no fundamental or philosophical difficulties to be encountered with the synthesis of the monomers. As will now be apparent, one can say nearly as much for the formation of the biological polymers.

The central problem in understanding how the polymers were formed on the primitive earth is understanding how reactions requiring both the input of energy and the removal of water could take place in the ocean. Each joining of subunits to lengthen a polymer chain calls for the removal of the elements of water from the ends being joined [*see equation 12 in bottom illustration on page 38*]. Since such reactions are reversible, an excess of water will drive them toward the left, in the direction of hydrolysis rather than polymerization. Furthermore, if all the reactants and products are present in comparable concentrations, the reaction to the left gives off free energy and hence is spontaneous, whereas the desired reaction requires free energy and hence must be driven "uphill" to the right. There are two ways to drive the polymerization reaction to the right: concentrate the reactants and remove water from the products or couple the process to some energy-releasing reaction that will drive polymerization toward completion. Both approaches have been investigated.

The energy to drive polymerization reactions in living organisms today is supplied by molecules of ATP. The coupling of reactions that require energy with those that release energy is accomplished by enzymes. On the prebiological earth, before enzymes existed, the two functions could have been carried out by certain compounds that possess large amounts of free energy and provide their own coupling to the reactant molecules. Such coupling reagents are familiar to organic chemists. The molecules of a typical class, the carbodiimides, have a carbon atom linked by energetic double bonds to two nitrogen atoms (N=C=N). If the carbodiimide is brought in contact successively with two monomers or polymers, *A* and *B*, one of which has a terminal hydroxyl group and the other a terminal hydrogen atom, it removes water and joins the two monomers or polymers end to end. The uptake of water by the carbodiimide releases enough energy to make the combined reaction go [*see equations 13 and 14 in bottom illustration on page 38*].

Carbodiimides are only illustrative of the principle of coupling. Potential coupling agents that have actually been made in prebiological synthesis experiments include cyanogen (N≡C–C≡N), cyanamide ($N{\equiv}C{-}NH_2$), cyanoacetylene (N≡C–C≡C–H) and diaminomaleonitrile, all of which incorporate carbon and nitrogen atoms joined by high-energy triple bonds. Cyanoacetylene can be produced by an electric discharge through a mixture of hydrogen cyanide; cyanogen can be produced from hydrogen cyanide both by electric discharge and by ultraviolet radiation. The energy of the electric spark or of the photons is stored as free chemical energy in the triple bonds of the product molecules for release later in the coupling reaction. In this way prebiological polymerizations could be driven indirectly by ultraviolet radiation or lightning in much the same way that animals living on plant starches today are driven indirectly by solar energy.

A major problem presented by such coupling mechanisms under prebiological conditions is to explain how the coupling agent can be prevented from combining directly with the ubiquitous water molecules and thereby short-circuiting the desired polymerization reactions. In the laboratory carbodiimide coupling reactions are carried out in nonaqueous solvents, but that is clearly not a reasonable model for the primitive earth. For example, if the primitive coupling agent were cyanogen, water could abort polymerization in either the first or the second step of the two-step reaction [*see equation 15 in bottom illustration on page 38*].

One suggestion is that prebiological coupling reactions could have succeeded in aqueous solution if the molecules to be polymerized had been previously coupled to negatively charged ions, such as the phosphate ion (HPO_4^{--}). Organophosphates can compete quite successfully with water for the energetic bonds of coupling reagents. Phosphate condensations have been used successfully to make dipeptides from amino acids, to make adenosine monophosphate from adenosine and phosphate, to connect the ribose-phosphate backbone of nucleic acids and to build up polyphosphates from phosphate ions.

This nonbiological synthesis of polyphosphates, long-chain polymers of phosphate, may have been of great significance in the evolution of life. ATP is useful to living cells as an energy-storage molecule precisely because the hydrolysis of one bond to produce adenosine diphosphate (ADP) and one unit of inorganic phosphate releases a large amount of chemical free energy. The energy comes in part from the repulsion between the negative charges on the molecular fragments of the reaction. As might be expected, comparably large amounts of energy are released when polyphosphates are hydrolyzed to phosphate [*see equations 16 and 17 in top illustration on preceding page*]. ATP can be regarded simply as a small polyphosphate with an adenine "label" attached so that it can be recognized by enzymes. Some present-day bacteria store energy in polyphosphate bodies within their cytoplasm. It may be that polyphosphates produced by condensing agents were the first energy source tapped by living organisms or their immediate precursors.

Glycolysis, or anaerobic fermentation, is the most universal and probably the oldest energy-extracting pathway found in life on the earth today. Its function is to break down glucose or similar molecules and store the energy in the form of "labeled polyphosphate": ATP. The glycolytic pathway may have arisen only in response to a shortage of natural polyphosphates, as the energy needs of a growing population of primitive organisms exceeded the natural rate of production of polyphosphates by condensation. If one assumes that coupling reagents were synthesized with the energy from ultraviolet radiation or electric discharge, if the coupling reagents were responsible for the buildup of polyphosphates and if hydrolysis of the polyphosphates provided the energy source for primitive life, then the first organisms would have lived on the energy of lightning and ultraviolet radiation at third hand.

The difficulty of preventing competition by water molecules in a coupling reaction has encouraged biochemists to think of ways in which the amount of water in the vicinity of the polymerizing species might be reduced. An obvious possibility is evaporation, and one can imagine a small portion of the Haldane soup being concentrated by the solar evaporation of a tide pool near some Archean beach. A freshwater pond would be even better, since salt would not crystallize as the water dried up. One objection to this proposal is that several of the important precursors of biological molecules, such as hydrogen cyanide, cyanogen, formaldehyde, acetaldehyde and ammonia, are themselves volatile. The evaporation of pools might be more effective in concentrating monomers for polymerization than in the synthesis of monomers themselves.

A possibly more attractive mechanism for concentrating prebiological molecules, first pointed out by Bernal, involves the adsorption of molecules on the surface of common minerals. Micas and clays, for example, consist of stacked silicate sheets held together by positive ions, with layers of water molecules between the sheets. The water layers make both sides of the silicate sheets accessible to molecules that diffuse into the clays, so that the total surface area available for adsorbing molecules is enormous. In kaolinite clay the silicate layers are separated by only .71 nanometer, which means that a cube of kaolinite one centimeter on a side pro-

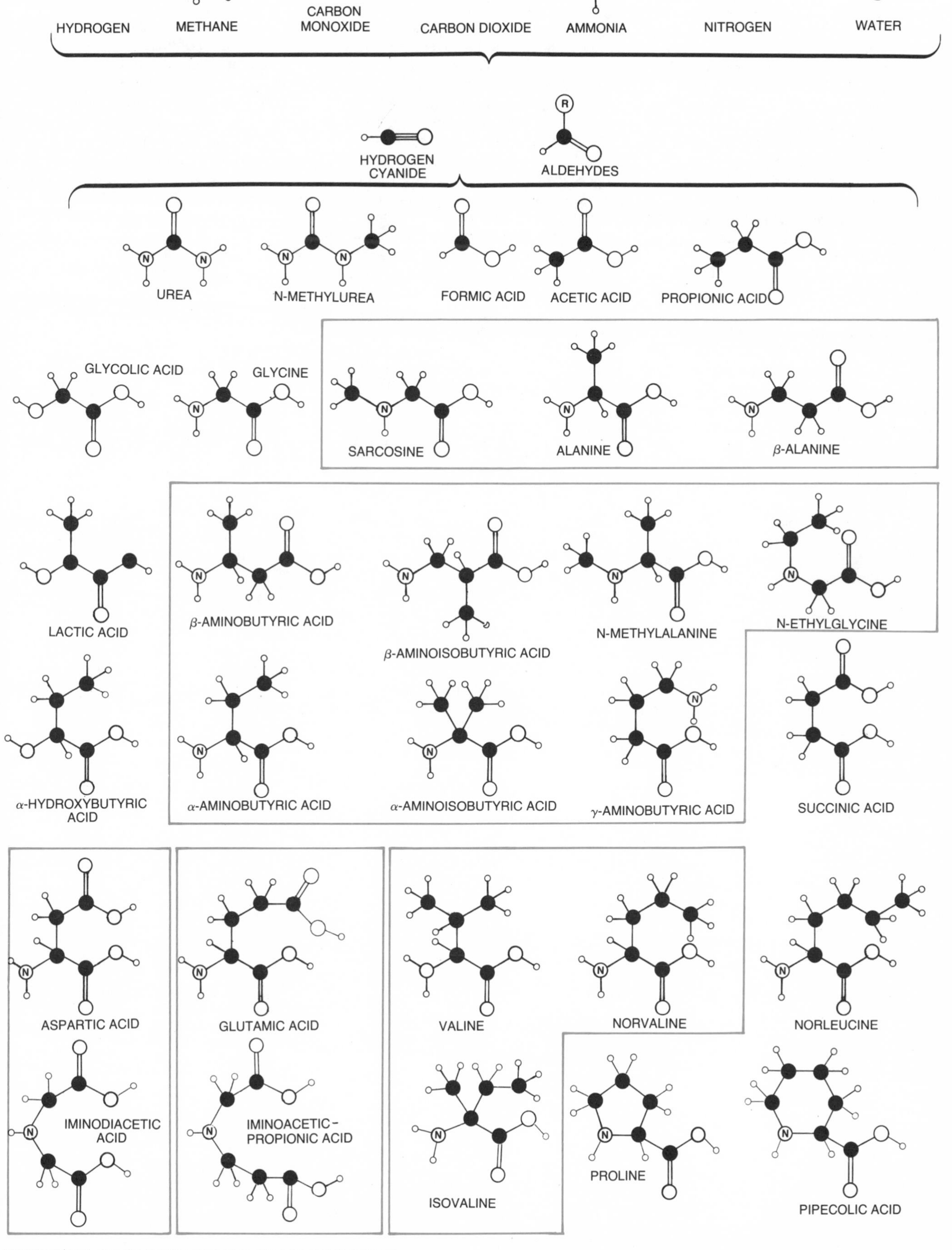

PRODUCTS OF SPARK-DISCHARGE EXPERIMENTS include many compounds found in living organisms and others that, although they are closely related, are not found in living matter today. The raw materials for the experiment are at the top. The first intermediates formed are hydrogen cyanide and a variety of aldehydes. The products are arranged so that the number of carbon atoms increases from left to right and from top to bottom. Isomers (molecules with the same atoms but different configurations) are enclosed by colored lines. These 30 typical products include 20 compounds listed in illustration at top of page 35. Six compounds are amino acids found in proteins.

vides a total surface area of about 2,800 square meters, or two-thirds the area of a football field. Furthermore, the silicate sheets themselves are negatively charged and the aluminum ions are bonded to the sheets by triple positive charges. The abundance of negative and positive charges not only can serve to bind charged molecules to the sheets but also can act as primitive catalytic centers for reactions.

Aharon Katchalsky of the Weizmann Institute of Science in Israel demonstrated that the montmorillonite clays will promote the polymerization of protein-like polypeptide chains from amino acid adenylates, which are esters formed from amino acids and adenosine monophosphate. Because they are rich in free energy and incorporate phosphate ions the adenylates polymerize efficiently even in aqueous solution [*see equation 18 in middle illustration on page 39*]. When adenylates are adsorbed on clay surfaces, they form polypeptide chains of 50 or more amino acids with nearly 100 percent efficiency. Amino acid adenylates are the precursors of protein synthesis in all living organisms, and so it is tempting to imagine that this clay-surface polymerization with the same precursors might be an early step in the evolution of biological protein synthesis. Once the polymers had been formed they could be leached back into solution to accumulate slowly over the aeons, ready for further reactions.

Two other means of concentration and polymerization of prebiological substances have been proposed: freezing and heating to dryness. Miller and Orgel have pointed out that solutions can be concentrated by freezing out the water as ice crystals, a procedure familiar to many as a means of making applejack from hard cider. On the prebiological earth the freezing of ice crystals out of a dilute solution of hydrogen cyanide could finally yield a solution containing 75 percent hydrogen cyanide by weight, freezing at minus 21 degrees Celsius.

At the other end of the temperature range, as Sidney W. Fox of the University of Miami has shown, dry mixtures of pure amino acids will polymerize spontaneously in a few hours at temperatures as low as 130 degrees C. to produce what Fox calls thermal proteinoids. If polyphosphates are present, similar results can be obtained by merely warming the amino acid mixture to 60 degrees for a day or so. Provided that the amino acids in the mixture are predominantly either acidic or basic and have side chains that are electrically charged, Fox's method will build polymers consisting of 200 or more amino acid units. Although most of the peptide bonds formed are of the normal type, a small fraction exhibit "wrong" connections involving the side chains. This is hardly surprising. One would not expect a prebiological polymer to show the degree of perfection found in a product of living metabolism. Fox speculates that amino acids formed in the ocean could have been washed up on volcanic cinder cones, evaporated to dryness and polymerized by heat. The resulting proteinoids, on being washed back into the sea, would have been available for further prebiological processing.

The problem of ensuring that only the right connections are made in nonbiological polymerization is rather more acute for nucleic acids than it is for proteins. As we have seen, each ribose molecule has four hydroxyl groups that can be involved in binding a purine or pyrimidine base and in polymerizing with bridging phosphates. Assuming that an efficient nonenzymatic method could be found for making nucleotides (a base plus a ribose plus a phosphate) with all the correct linkages, there would still be the problem of joining the nucleotides correctly to make polymers of nucleic acid. Although nucleotides can be polymerized nonbiologically with mild heat (about 55 degrees C.) in the presence of polyphosphates, the most readily formed connection is from the 5′ hydroxyl of one sugar to the 2′ hydroxyl of the next sugar rather than to the 3′ hydroxyl, the connection that is found in all DNA's and RNA's today. The 5′,3′ linkage must have had a significant advantage over the 5′,2′ one to have been adopted for storage of genetic information even though it is less favored chemically.

Studies with molecular models show that it is possible to construct a double-strand DNA helix with paired bases and a 5′,2′ connection, but the helix appears to be less stable than one with a 5′,3′ structure. Hence a genetic message stored in a 5′,2′ helix may have been less secure than a message stored in a 5′,3′ helix. One way of ensuring that a 5′,2′ helix does not form is to remove the 2′ hydroxyl group, and that is exactly what makes DNA different from RNA. DNA may be the more primitive of the two information-storing polymers, with RNA appearing only after enzymes had been developed that would avoid making the connection to the 2′ hydroxyl group.

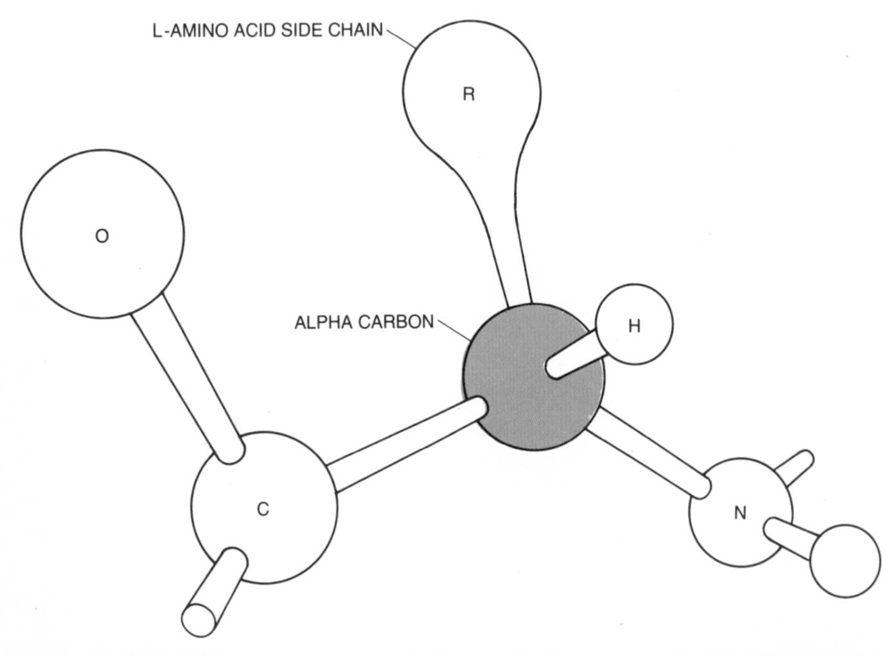

STEREOISOMERS are molecules that have two configurations, one the mirror image of the other. Among organic molecules stereoisomers can be formed when a carbon atom in the molecule has four different atoms or groups of atoms attached to it. Such a carbon atom is termed asymmetric. The central, or alpha-carbon, atom (*color*) in an amino acid is asymmetric. An amino acid whose side chain (*R*) projects to the left as one crosses an imaginary bridge from the carbonyl group (CO) to alpha carbon to nitrogen atom (N) is an L-amino acid. In a D-amino acid side chain *R* projects to right. Living organisms make proteins only from L-amino acids.

Living organisms that share an environment with other organisms must be clearly set off from that environment by a boundary surface to avoid being diluted out of existence. The segregation of matter in solution into droplets that were possible precursors of life has been studied mainly by two men and their coworkers: Oparin and Fox. Oparin has focused for many years on the tendency of aqueous solutions of polymers to separate spontaneously into coacervates: polymer-rich colloidal droplets suspended in a water-rich surrounding medium. Various combinations of biological polymers will give rise to coacervates: protein-carbohydrate (histone and gum arabic), protein-protein (histone and albumin) and protein-nucleic acid (histone or clupein with DNA or RNA). Such coacervates are to be regarded not as ancestors of living cells, since the polymers employed by Oparin in his experiments are definitely not primitive, but rather as analogues of the kinds of complex chemical behavior that can arise under the influence of natural forces.

The coacervate droplets range in diameter from one micrometer to 500 micrometers. Many seem to be set off from the surrounding medium by a kind of membrane, a thickening around the outside of the droplet of the polymer that causes it to separate from the bulk medium in the first place. Some coacervate systems are unstable: the droplets settle to the bottom of the liquid within minutes and coalesce into a nonaqueous layer. Oparin and his co-workers have sought conditions that will stabilize the suspensions of coacervate droplets for hours or weeks. Interestingly enough, they have found that one way to stabilize the droplets is to give them a primitive kind of metabolism.

One important property of coacervates, or of any two-phase system, is that substances whose solubility differs in the two phases will be preferentially concentrated in one phase or the other. Oparin found that when he added the enzyme phosphorylase to a solution containing histone and gum arabic, the enzyme was concentrated within the coacervate droplets. If glucose-1-phosphate was then added to the surrounding water, it diffused into the droplets and was polymerized to starch by the enzyme. Since gum arabic itself is a sugar polymer, the starch adds to the bulk of the droplet, causing it to grow in size. Energy for polymerization comes from the phosphate bond in glucose-1-phosphate. The inorganic phosphate that is released diffuses back out of the droplet into solution as a waste product.

When the coacervate droplets get too big, they tend to break up spontaneously into several daughter droplets. Those that happen to receive molecules of phosphorylase enzyme can continue to grow, although they do so at a lower rate because the original supply of enzyme molecules is dispersed among many droplets. If there were some way for the droplets to make more phosphorylase molecules (and it is a very big if), such coacervates would be examples of self-perpetuating protoorganisms with a one-step energy metabolism. They would be able to survive, grow and multiply on a restricted diet of glucose-1-phosphate.

If both phosphorylase and amylase are added to the coacervate preparation, both enzymes accumulate within the droplets and a two-step reaction ensues. Glucose-1-phosphate diffuses into the droplets and is polymerized to starch by phosphorylase. Amylase then cuts the starch polymer down to maltose, a dimer of glucose. The maltose diffuses back into the bulk solution along with the inorganic phosphate. The coacervates are thus small factories, driven

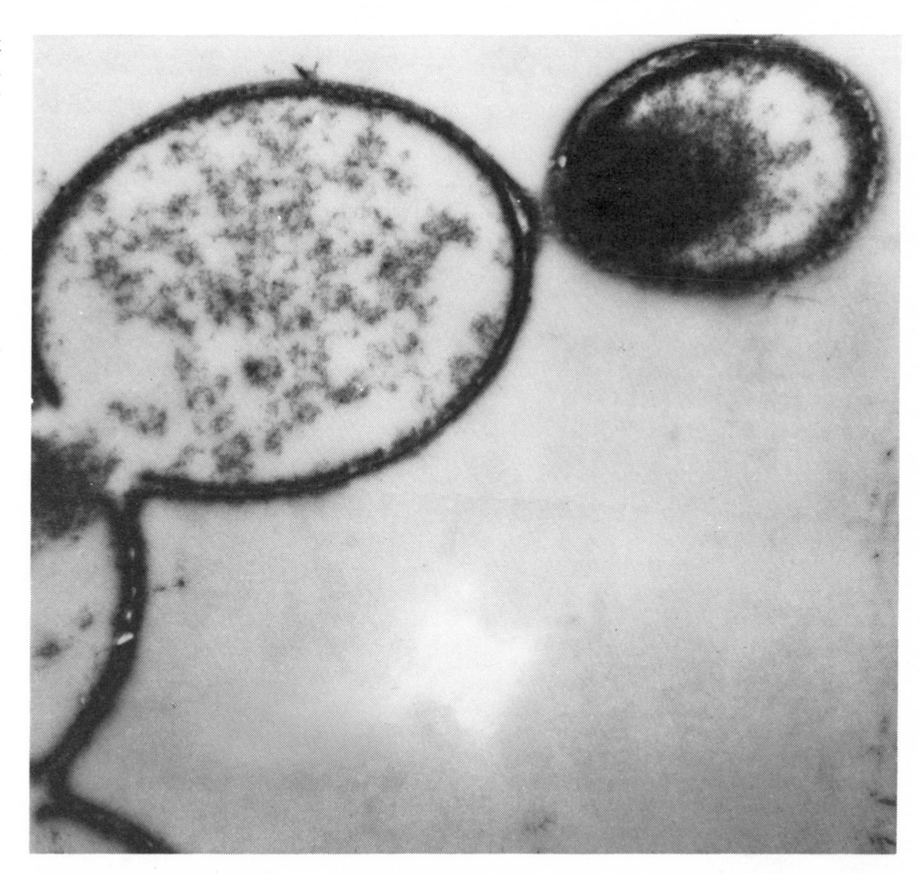

COACERVATES, polymer-rich colloidal droplets, have been studied in the Moscow laboratory of A. I. Oparin because of their conjectural resemblance to prebiological entities. These coacervates are droplets formed in an aqueous solution of protamine and polyadenylic acid. Oparin has found that droplets survive longer if they can carry out polymerization reactions.

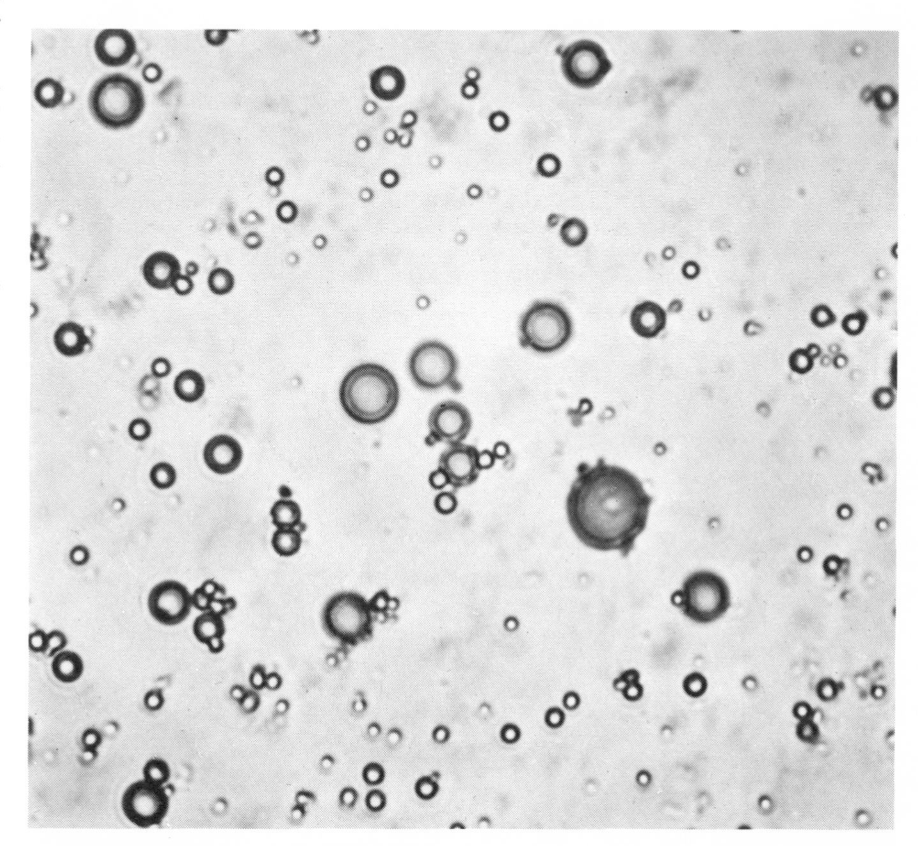

ANOTHER KIND OF MICROSPHEROIDAL AGGREGATE, studied by Sidney W. Fox of the University of Miami, forms from "thermal proteinoid," a polymer produced by heating dry mixtures of amino acids to moderate temperatures. Under suitable conditions thermal proteinoid will form microspheres several micrometers in diameter, which grow slowly and eventually bud. The microspheres seem to have a two-layer membrane suggestive of that in bacteria.

by the energy of the glucose-phosphate bond, for dimerizing glucose-1-phosphate to maltose.

Oparin has reported another self-growing system in which the coacervate droplets are made from histone and RNA. The enzyme RNA polymerase is introduced into the droplets, and ADP is added to the surrounding medium as "food." When the ADP enters the droplet, it encounters the RNA polymerase and is polymerized into polyadenylic acid. The energy for polymerization is contained within the ADP itself. The new polyadenylic acid adds to the total RNA in the coacervates. The droplets grow with time and break up into daughter droplets. Such systems eventually wind down because the supply of enzyme molecules for polymerizing ADP does not increase with the total mass of the coacervate droplets. As we saw earlier, however, nucleic acids can be polymerized nonenzymatically with small, energy-rich coupling-agent molecules such as cyanogen. It should be possible to construct coacervate droplets from protein and RNA, to provide them with ADP and the appropriate coupling reagents, and to see them grow and multiply without limit as long as their "nutrients" continue to be supplied.

Oparin has also set up coacervate-droplet experiments that mimic electron transport. The droplets contain a dehydrogenase enzyme from bacteria: nicotinamide adenine dinucleotide dehydrogenase (NADH). NADH and the dye methyl red are added to the medium and diffuse into the droplets. At the active site on the enzyme the NADH gives up hydrogen, which reduces the dye. The reduced dye and the oxidized NAD^+ diffuse out of the droplets.

In another dye-reduction experiment, which mimics photosynthesis, chlorophyll is incorporated into the droplets; methyl red and ascorbic acid are added to the surroundings as nutrients. Ascorbic acid by itself is not a strong enough reducing agent to reduce methyl red. If, however, the droplets are illuminated with visible light, excited electrons from the chlorophyll can reduce the methyl red, and the electrons can be replaced on the chlorophyll by taking them away from ascorbic acid. In this way ascorbic acid, assisted by the energy of the photons of light, can reduce methyl red in a process that is analogous to the way water molecules, when they are assisted by photon energy, can reduce $NADP^+$ to NADPH in the photosynthesis conducted by green plants.

Fox's interest in coacervate-like droplets has developed from his work with thermal proteinoids. The proteinoids have a remarkable property: when they are heated in a concentrated aqueous solution at 130 to 180 degrees C., they aggregate spontaneously into microspheres one or two micrometers in diameter. Although no lipids are present, many of the microspheres develop an outer boundary that resembles the double lipid layer of a cell membrane. Under the proper conditions the microspheres will grow at the expense of the dissolved proteinoid and will bud and fission in a most bacteriumlike manner.

Whereas Oparin has constructed artificial systems with catalysts incorporated, Fox has looked for catalytic activity inherent in the microspheres themselves. For example, he has found that microsphere preparations can catalyze the decomposition of glucose and can function as esterases and peroxidases. It would be surprising indeed if a polypeptide chain with positive and negative charges on its side groups did not exhibit some kind of generalized acid-base catalytic activity. Perhaps specific enzymes evolved from such randomly ordered polymers by a gradual improvement in the positioning of electron-donating and electron-accepting side chains at active sites that were tailored to favor one reaction over another.

The Oparin and Fox experiments are only analogies to life, but they are suggestive ones. They demonstrate the extent to which lifelike behavior is grounded in physical chemistry, and they illustrate the concept of chemical selection for survival. This is the only kind of natural selection and evolution that could have existed prior to the de-

POLYSACCHARIDE MEMBRANE OR WALL
STARCH
GLUCOSE-1-PHOSPHATE IN
PHOSPHATE OUT
PHOSPHORYLASE
COACERVATE DROP

$$n(\text{GLUCOSE-PHOSPHATE}) \xrightarrow{\text{PHOSPHORYLASE}} (\text{GLUCOSE})_n + {}_nHPO_4^{--}$$

GLUCOSE-1-PHOSPHATE — STARCH — PHOSPHATE

POLYMERIZATION INSIDE A COACERVATE DROPLET causes the wall of the droplet to thicken and the droplet to grow. The droplet, consisting of protein and polysaccharide, contains the enzyme phosphorylase. Glucose-1-phosphate diffuses into the droplet and is polymerized to starch by the enzyme. The starch migrates to the wall and increases volume of droplet.

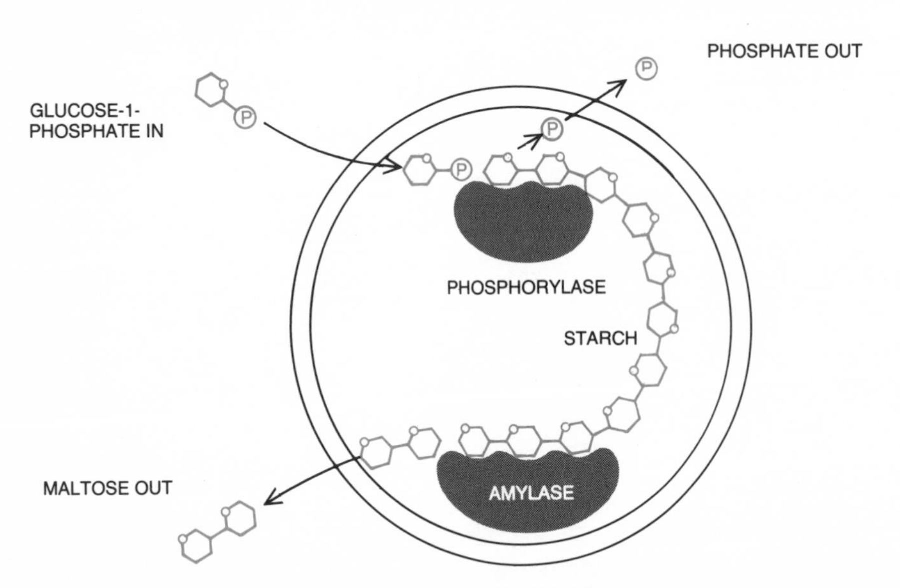

TWO-STEP REACTION takes place inside a protein-carbohydrate droplet provided with two enzymes. One enzyme, phosphorylase, polymerizes glucose-1-phosphate to starch. The second enzyme, amylase, degrades the starch to maltose. Droplets in this instance do not grow because the starch disappears as fast as it is made. The maltose diffuses back into surrounding medium.

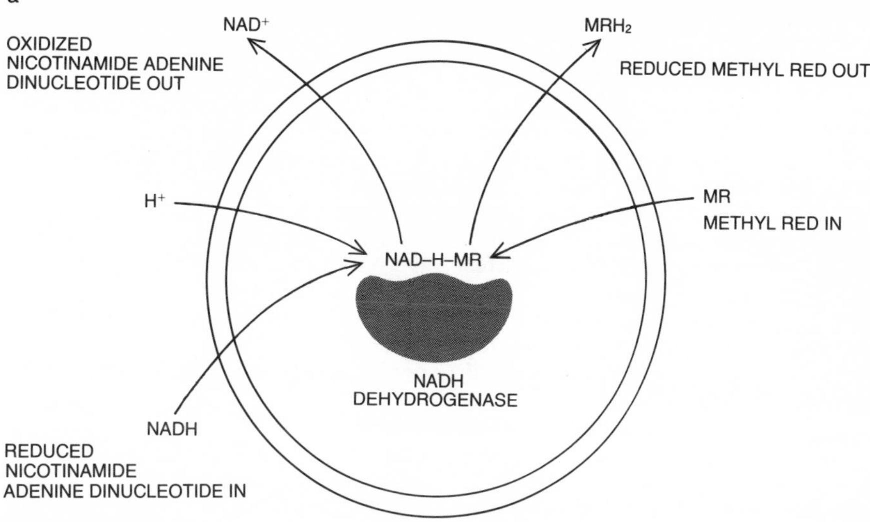

ELECTRON TRANSPORT IS MIMICKED in an Oparin experiment in which coacervate droplets are supplied with NADH dehydrogenase, an enzyme from bacteria. The medium contains methyl red, a dye, and nicotinamide adenine dinucleotide in its reduced form (NADH). When the two substances diffuse into the droplet, the enzyme effectuates the transfer of hydrogen from NADH to the dye, thus reducing it. Products of reaction then diffuse out of droplet.

velopment of information-storing molecules and genetic selection. Such experiments demonstrate that separation into coacervate suspensions or microspheres is a common behavior of polymers in solution, that all such microsystems are not equally stable and that the probability of their survival is enhanced if they have within themselves the ability to carry out simple reactions increasing their bulk or strengthening their barrier against the outside world.

One can visualize that before living cells evolved the primitive ocean was teeming with droplets possessing special chemistries that survived for a time and then were dispersed again. Those droplets that by sheer chance contained catalysts able to induce "useful" polymerizations would survive longer than others; the probability of survival would be directly linked to the complexity and effectiveness of their "metabolism." Over the aeons there would be a strong chemical selection for the types of droplets that contained within themselves the ability to take molecules and energy from their surroundings and incorporate them into substances that would promote the survival not only of the parent droplets but also of the daughter droplets into which the parents were dispersed when they became too big. This is not life, but it is getting close to it. The missing ingredient is an orderly mechanism for ensuring that all the daughter droplets receive the catalysts they need for all the reactions important to their survival. This is the pragmatic definition of a genetic apparatus, the subject to which I shall now turn.

The evolution of the genetic machinery is the step for which there are no laboratory models; hence one can speculate endlessly, unfettered by inconvenient facts. The complex genetic apparatus in present-day organisms is so universal that one has few clues as to what the apparatus may have looked like in its most primitive form.

Some 30 years ago Norman H. Horowitz of the California Institute of Technology made the provocative suggestion that metabolic systems evolved, so to speak, from back to front. If today a series of metabolic steps goes from substance *A* to substance *B* and then to substances *C*, *D* and *E*, the oldest need was probably for substance *E*, and the oldest reaction was the one that made *E* from *D*, which then was a raw material obtained from the surroundings. Only when the supply of *D* began to run low would there have been a strong selection pressure for the ability to make *D* from another raw material, *C*. An eventual shortage of *C* would have led to competition in finding ways of making it from some other precursor, *B*, and in this manner an entire metabolic chain could have evolved slowly in reverse order.

In this way of thinking photosynthesis evolved as a means of providing an alternate source of glucose for organisms that depended on anaerobic fermentation of compounds rich in free energy, at a time when competition had depleted the natural supply of such compounds. Anaerobic fermentation, or glycolysis, itself may have evolved as an alternate means of providing a supply of ATP for more primitive organisms that previously had been dependent on an external source of nucleotides and polyphosphates for driving energy. Hence the familiar metabolic series of reactions (1) photosynthesis of glucose, (2) glycolysis with energy storage through ATP and (3) the utilization of ATP as the energy source in cell activities may be the result of the kind of back-to-front evolution Horowitz postulated.

At a still earlier stage enzymes themselves may not have been essential if a plentiful supply of activated monomers and condensing reagents was available. Enzymes not only are catalysts but also have a directing or coupling function, ensuring that the chemical free energy released by one reaction is utilized productively by another reaction rather than being dissipated as heat. As soon as it became important for a limited supply of free energy to be channeled into one or a few of the many possible reactions, directed catalysis by enzymes would have become essential.

The first protoenzymes may have been the polymer chains that were themselves being formed. Some polymerizations tend to be autocatalytic: the presence of a particular polymer favors the formation of more of the same polymer. The double-strand helix of DNA is an outstanding example of autocatalytic polymerization, and for this reason alone DNA may have been the natural candidate for a central role in living organisms. As soon as the reaction ceases to be strictly autocatalytic, that is, when the catalyst for the reaction is no longer just the product of the reaction, the problem arises of ensuring that the supply of catalyst increases and is passed on to the descendants of the protobiont.

The first successfully stabilized protobionts may have been autocatalytic coacervates of nucleic acids similar to those constructed by Oparin but dependent on activated monomers and coupling reagents rather than on polymerase enzymes. If the nucleic acid could have served as a template for the polymerization of protein chains, even of a random sequence, then this protein might have been useful as a skin to protect the nucleic acid coacervate. In this way a cooperative interaction between nucleic acid and protein would have existed from the beginning, with the nucleic acid playing the autocatalytic and template role, and the protein playing a structural and protective role. If a particular pattern of positive and negative charges along the polypeptide chain proved to be helpful in polymerizing either the nucleic acid or the polypeptide, then the first protein catalyst or enzyme function would have arisen. There would then have been strong selection pressures for those nucleic acid sequences that favored the continued formation of just that pattern of positive and negative amino acid side chains. In this way template replication in the nucleic acid and enzymatic catalysis in the polypeptide could have evolved in tandem, and there may never have been an era either of "life without DNA" or of "naked genes."

Such speculations all require the existence of some kind of mutual recognition or complementarity at the molecular level between amino acid sequences in proteins and base sequences in nucleic acids. Many attempts have been made to find a natural fit between protein sequences and nucleic acid sequences that could have existed before the appearance of the present-day elaborate machinery involving transfer-RNA molecules, ribosomes and charging enzymes. None of these attempts has been fully convincing. In all present-day life a charging enzyme attaches a specific amino acid to a transfer-RNA molecule that has at its other end an anticodon for that amino acid. (An anticodon is a triplet of bases that are complementary to a codon: a triplet that codes for a particular amino acid.) The specificity of matching amino acids to codons lies neither in the codon nor in the transfer RNA but in the charging enzyme. How did the matching arise before charging enzymes existed? This looks like another chicken-and-egg paradox, since the charging enzymes themselves are synthesized by the translation machinery they help to operate. The answer to the original chicken-and-egg paradox was that neither the chicken nor the egg came first; they evolved together from lower forms of life. The same must be true of the genetic machinery; the entire apparatus evolved in concert from simpler systems now driven out of existence by competition. Although we can examine fossil remains of chicken ancestors, we have no fossil enzymes to study. We can only imagine what probably existed, and our imagination so far has not been very helpful.

The system today that is most likely to shed light on a primitive association between nucleic acid and the replication of protein is the repressor-operator system of genetic control. Although the direct recognition of nucleic acid sequences by amino acid side chains is no longer a part of the readout of the genetic message, when certain genes are shut down in bacteria, a protein molecule of definite amino acid sequence (the repressor) must recognize and bind to a particular sequence of base pairs (the operator DNA). The nature of this sequence recognition is being studied in many laboratories. When it is finally understood exactly how the protein repressor recognizes the base sequence of the DNA operator, we may begin to hypothesize intelligently how a given sequence of bases could have produced a specific polypeptide chain sequence in the days before transfer RNA, ribosomes and charging enzymes.

Through some gradual means about which we can now only speculate, an association of nucleic acid as the archival material with protein as the working catalysts evolved into the complex genetic transcription and translation machinery that all forms of life exhibit today. This made it possible to preserve all the biochemical abilities of a parent protocell in its offspring. But since the genetic message was subject to alteration by the slow accumulation of errors and by direct mutation brought about by ionizing radiation and other agents, the environment could now serve as a screen, selecting for or against the possessors of the altered messages. Evolution by natural selection in its Darwinian sense could begin.

The steps I have outlined so far or something similar to them are probably responsible for the appearance of the first living organisms on the earth. They were presumably one-celled entities resembling modern fermenting bacteria such as *Clostridium,* which had a complete genetic apparatus but were totally dependent on the breakdown of nonbiologically formed energy-rich molecules for their survival. They would have been scavengers of the organic matter produced by electric discharges and ultraviolet radiation. Hence the total amount of life the earth could have sustained would have been limited by the rate of production of such compounds by nonbiological means. Living organisms in that era would have been strictly consumers of organic matter, not producers.

The capacity of the earth to support life was enormously enhanced by the invention of photosynthesis, which enabled living organisms to capture solar energy for the synthesis of organic molecules. The first photosynthesizers removed themselves from the competition for a dwindling supply of natural energy-rich molecules and set themselves up as primary producers. Photosynthesis using hydrogen sulfide as the source of hydrogen atoms for reducing carbon dioxide, which is the process conducted today by the green and purple sulfur bacteria, undoubtedly preceded the more elaborate two-step form of photosynthesis wherein water supplies the hydrogen, which is the process conducted today by the cyanobacteria, or blue-green algae, and by green plants [*see equations 19 and 20 in bottom illustration on page 39*]. On the primitive earth hydrogen sulfide would have been sufficiently abundant for it to have served as a practical reductant. Water is even more abundant, however, and organisms that found ways of taking hydrogen atoms for synthesis from water rather than from hydrogen sulfide would have had a great advantage over their sulfur-using cousins.

This brings the story of life on the earth up to the cyanobacteria, whose fossilized ancestors seem to be present in sediments at least 3.2 billion years old. It is apparent that not only life but also photosynthetic life evolved within a billion years of the formation of the planet. It is not absolutely clear that those ancient organisms split water by photosynthesis and released free oxygen into the atmosphere, but it seems likely.

The next two billion years saw a revolution in the nature of the atmosphere of the planet: from a reducing atmosphere with little or no free oxygen to an oxidizing atmosphere in which one out of every five molecules is oxygen. One consequence was the formation of an ozone layer in the upper atmosphere that sharply reduced the ultraviolet radiation at the earth's surface. Although this effectively ended the nonbiological synthesis of organic matter, biological photosynthesis working with the energy in the visible wavelengths more than made up for it. The pattern of life driven by solar energy was fixed for all time on our planet, and the stage was set for true biological evolution.

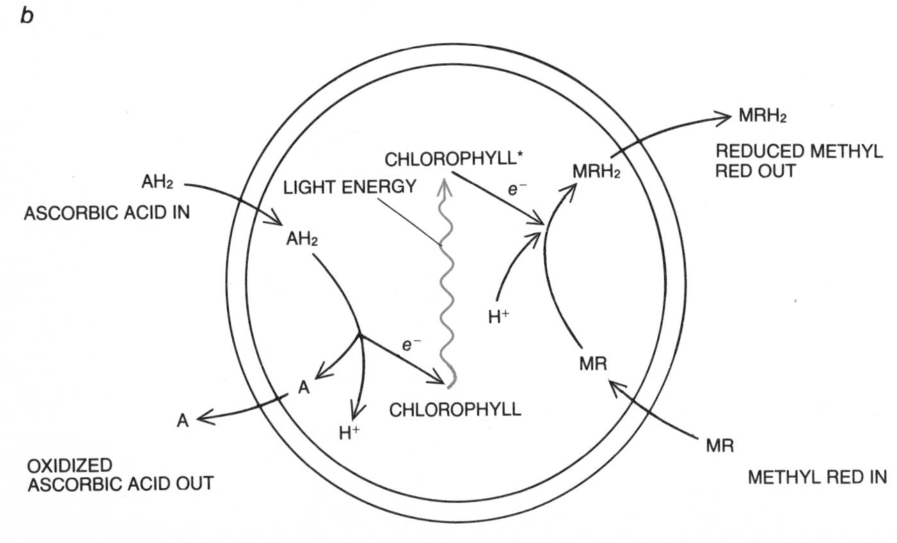

PHOTOSYNTHESIS IS MIMICKED in another Oparin experiment with coacervates containing chlorophyll. Here again the dye methyl red is the substance to be reduced. Ascorbic acid, which diffuses into the droplets, is not in itself a strong enough reducing agent to serve the purpose. If the droplets are exposed to light, however, excited electrons from the chlorophyll are capable of reducing the methyl red. The chlorophyll then regains the expended electrons from the ascorbic acid, which is oxidized in the process and diffuses out of the droplets.

IV

The Evolution of the Earliest Cells

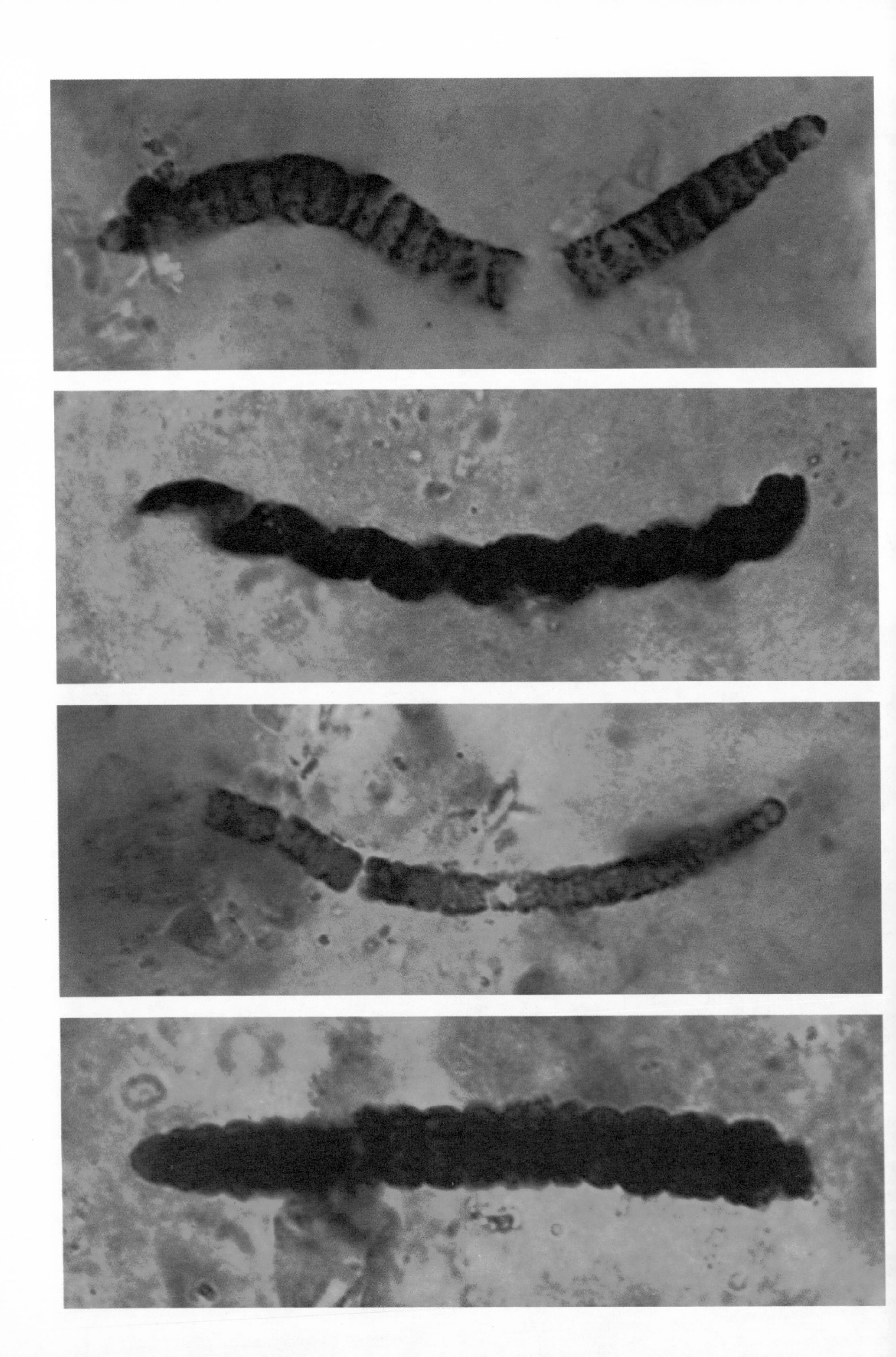

The Evolution of the Earliest Cells

BY J. WILLIAM SCHOPF

For some three billion years the only living things were primitive microorganisms. These early cells gave rise to biochemical systems and the oxygen-enriched atmosphere on which modern life depends

When *On the Origin of Species* appeared in 1859, the history of life could be traced back to the beginning of the Cambrian period of geologic time, to the earliest recognized fossils, forms that are now known to have lived more than 500 million years ago. A far longer prehistory of life has since been discovered: it extends back through geologic time almost three billion years more. During most of that long Precambrian interval the only inhabitants of the earth were simple microscopic organisms, many of them comparable in size and complexity to modern bacteria. The conditions under which these organisms lived differed greatly from those prevailing today, but the mechanisms of evolution were the same. Genetic variations made some individuals better fitted than others to survive and to reproduce in a given environment, and so the heritable traits of the better-adapted organisms were more often represented in succeeding generations. The emergence of new forms of life through this principle of natural selection worked great changes in turn on the physical environment, thereby altering the conditions of evolution.

One momentous event in Precambrian evolution was the development of the biochemical apparatus of oxygen-generating photosynthesis. Oxygen released as a by-product of photosynthesis accumulated in the atmosphere and effected a new cycle of biological adaptation. The first organisms to evolve in response to this environmental change could merely tolerate oxygen; later cells could actively employ oxygen in metabolism and were thereby enabled to extract more energy from foodstuff.

A second important episode in Precambrian history led to the emergence of a new kind of cell, in which the genetic material is aggregated in a distinct nucleus and is bounded by a membrane. Such nucleated cells are more highly organized than those without nuclei. What is most important, only nucleated cells are capable of advanced sexual reproduction, the process whereby the genetic variations of the parents can be passed on to the offspring in new combinations. Because sexual reproduction allows novel adaptations to spread quickly through a population its development accelerated the pace of evolutionary change. The large, complex, multicellular forms of life that have appeared and quickly diversified since the beginning of the Cambrian period are without exception made up of nucleated cells.

The history of life in its later phases, since the start of the Cambrian period, has been reconstructed mainly from the study of fossils preserved in sedimentary rocks. In the 18th and 19th centuries it gradually became apparent that the fossil record has appreciable chronological and geographical continuity. The fossil deposits form recognizable layers, which can be identified in widely separated geological formations. Boundaries between such layers, where one characteristic suite of fossils gives way to another, provide the basis for dividing geologic time into eras, periods and epochs.

One of the most dramatic boundaries in the rock record is the one that separates the Cambrian period from all that came before. The 11 periods of geologic time since the start of the Cambrian are referred to collectively as the Phanerozoic era, which might be translated from the Greek as the era of manifest life. The preceding era is called simply the Precambrian.

By itself the geologic time scale cannot provide dates for fossil deposits; it only lists their sequence. Ages can be calculated, however, from the constant rate of decay of radioactive isotopes in the earth's crust. By determining how much of an isotope has decayed since the minerals in a rock unit crystallized, a date can be assigned to that unit and to nearby strata containing fossils. Radioactive-isotope studies of this kind, carried out on rocks from many parts of the world, have established a rather well-defined date for the start of the Phanerozoic era: it began about 570 million years ago. The same method indicates that the earth itself and the rest of the solar system are 4.6 billion years old. Thus the Precambrian era encompasses some seven-eighths of the earth's entire history.

The boundary between the Precambrian era and the Cambrian period has traditionally been viewed as a sharp discontinuity. In Cambrian strata there are abundant fossils of marine plants and animals: seaweeds, worms, sponges, mollusks, lampshells and, what are perhaps most characteristic of the period, the early arthropods called trilobites. It was thought for many years that fossils were entirely absent in the underlying Precambrian strata. The Cambrian fauna seemed to come into existence abruptly and without known predecessors.

Life could not have begun with organisms as complex as trilobites. In *On the Origin of Species* Darwin wrote: "To the question why we do not find rich fossiliferous deposits belonging to . . . periods prior to the Cambrian system, I can give no satisfactory answer. . . . The case at present must remain inexplicable; and

MICROSCOPIC FOSSILS on the opposite page are the remains of organisms that were once the dominant form of life on the earth. The fossils are from silica-rich rocks in the Bitter Springs formation of central Australia, deposited about 850 million years ago, or late in the Precambrian era. The rocks have the layered structure of stromatolites, sedimentary deposits that were formed by matlike communities of microorganisms. Among Precambrian fossils these specimens are exceptionally well preserved; their petrified cell walls are composed of organic matter and have retained their three-dimensional form. In size, structure and ecological setting they resemble living cyanobacteria, or blue-green algae. Like their modern counterparts, the fossil forms were presumably capable of photosynthesis, and similar cyanobacteria some billion years earlier were evidently responsible for the first rapid release of oxygen into the earth's atmosphere. Organisms in these photomicrographs are about 60 micrometers long.

may be truly urged as a valid argument against the views here entertained." The argument is no longer valid, but it is only in the past 20 years or so that a definitive answer to it has been found.

One part of the answer lies in the discovery of primitive fossil animals in rocks below the earliest Cambrian strata. The fossils include the remains of jellyfishes, various kinds of worms and possibly sponges, and they make up a fauna quite distinct from that of the predominantly shelled animals of the Cambrian period. These discoveries, however, extend the fossil record by only about 100 million years, less than four percent of the Precambrian era. It can still be asked: What came before?

Since the 1950's a far-reaching explanation has emerged. It has come to be recognized that not only are many Precambrian rocks fossil-bearing but also Precambrian fossils can be found even in some of the most ancient sedimentary deposits known. These fossils had escaped notice earlier largely because they are the remains only of microscopic forms of life.

An important clue in the search for Precambrian life was discovered in the early years of the 20th century, but its significance was not fully appreciated until much later. The clue came in the form of masses of thinly layered limestone rock discovered by Charles Doolittle Walcott in Precambrian strata from western North America. Walcott found numerous moundlike or pillarlike structures made up of many draped horizontal layers, like tall stacks of pancakes. These structures are now called stromatolites, from the Greek *stroma,* meaning bed or coverlet, and *lithos,* meaning stone.

Walcott interpreted the stromatolites as being fossilized reefs that had probably been formed by various types of algae. Other workers were skeptical, and for many years the stromatolites were widely attributed to some nonbiological origin. The first convincing evidence substantiating Walcott's hypothesis came in 1954, when Stanley A. Tyler of the University of Wisconsin and Elso S. Barghoorn of Harvard University reported the discovery of fossil microscopic plants in an outcropping of Precambrian rocks called the Gunflint Iron formation near Lake Superior in Ontario. Most of the Gunflint fossils, which form the layers of dome-shaped and pillarlike stromatolites, resemble modern blue-green algae and bacteria. More recently, living stromatolites have been identified in several coastal habitats, most notably in a lagoon at Shark Bay on the western coast of Australia. They are indeed built up by communities of blue-green algae and bacteria, and they are strikingly similar in form to the fossilized Precambrian structures.

Today microfossils have been identified in some 45 stromatolitic deposits. (All but three of these fossilized communities have been found in the past 10 years.) The fossils are often well preserved, the cell walls being petrified in three-dimensional form, and they have become a prime source of documentation for the early history of life. In recent years the search for Precambrian microfossils in other kinds of sediments, such as shales deposited in offshore environments, has also been rewarded. These fossils are generally not as well preserved as the ones in stromatolites, most of them having been flattened by pressure; on the other hand, they supply information about Precambrian life in a habitat quite different from that of the shallow-water stromatolites.

A surprising amount of information can be derived from the fossil remains of a microorganism. Size, shape and degree of morphological complexity are among the most easily recognized features, but under favorable circumstances even details of the internal structure of cells can be discerned. In retracing the course of Precambrian evolution, however, there is no need to rely exclusively on the fossil record. An entirely independent archive has been preserved in the metabolism and the biochemical pathways of modern, living cells. No living organism is biochemically identical with its Precambri-

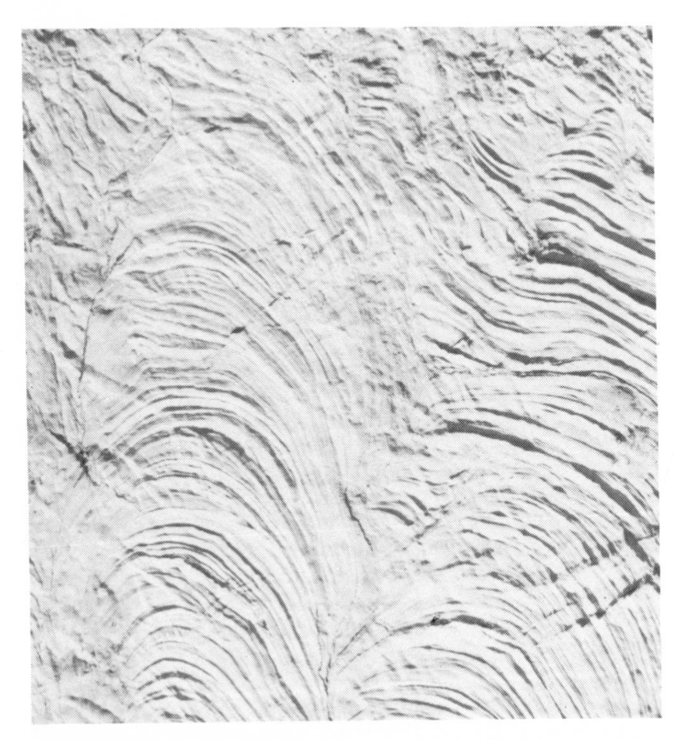

FOSSIL STROMATOLITES typically exhibit the appearance of mounds or pillars made up of many thin layers piled one on top of another. The stromatolites were formed by communities of cyanobacteria and other prokaryotes (cells without a nucleus) in shallow water; each layer represents a stage in the growth of the community. Stromatolites formed throughout much of the Precambrian era. They are an important source of Precambrian fossils. These specimens are in limestone about 1,300 million years old in Glacier National Park.

LIVING STROMATOLITES were photographed at Shark Bay in Australia. Elsewhere stromatolites are rare because of grazing by invertebrates. Here the invertebrates are excluded because the water is too salty for them; in the Precambrian era they had not yet evolved. In size and form the modern stromatolites are much like the fossil structures, and they are produced by the growth of cyanobacteria and other prokaryotes in matlike communities. The discovery of such living stromatolites has confirmed the biological origin of the fossil ones

an antecedents, but vestiges of earlier biochemistries have been retained. By studying their distribution in modern forms of life it is sometimes possible to deduce when certain biochemical capabilities first appeared in the evolutionary sequence.

Still another independent source of information about the early evolutionary progression is based neither on living nor on fossil organisms but on the inorganic geological record. The nature of the minerals found there reflects physical conditions at the time the minerals were deposited, conditions that may have been influenced by biological innovations. In order to understand the introduction of oxygen into the early atmosphere, for example, all three fields of study must be called on to testify. The mineral record tells when the change took place, the fossil record reveals the organisms responsible and the distribution of biochemical capabilities among modern organisms puts the development in its proper evolutionary context.

Since the 1960's it has become apparent that the greatest division among living organisms is not between plants and animals but between organisms whose cells have nuclei and those that lack nuclei. In terms of biochemistry, metabolism, genetics and intracellular organization, plants and animals are very similar; all such higher organisms, however, are quite different in these features from bacteria and blue-green algae, the principal types of non-nucleated life. Recognition of this discontinuity has been important for understanding the early stages of biological history.

Organisms whose cells have nuclei are called eukaryotes, from the Greek roots *eu-*, meaning well or true, and *karyon*, meaning kernel or nut. Cells without nuclei are prokaryotes, the prefix *pro-* meaning before. All green plants and all animals are eukaryotes. So are the fungi, including the molds and the yeasts, and protists such as *Paramecium* and *Euglena*. The prokaryotes include only two groups of organisms, the bacteria and the blue-green algae. The latter produce oxygen through photosynthesis like other algae and higher plants, but they have much stronger affinities with the bacteria than they do with eukaryotic forms of life. I shall therefore refer to blue-green algae by an alternative and more descriptive name, the cyanobacteria.

Several important traits distinguish eukaryotes from prokaryotes. In the nucleus of a eukaryotic cell the DNA is organized in chromosomes and is enclosed by an intracellular membrane; many prokaryotes have only a single loop of DNA, which is loose in the cytoplasm of the cell. Prokaryotes reproduce asexually by the comparatively simple process of binary fission. In contrast, asexual reproduction in eukaryot-

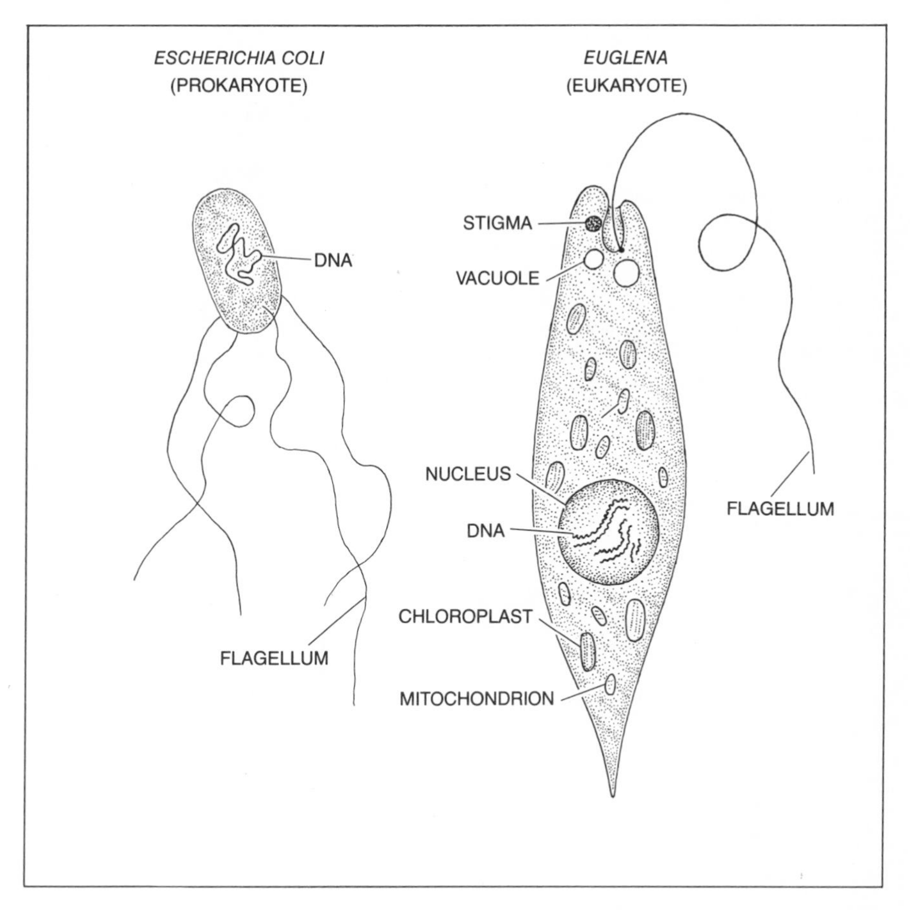

	PROKARYOTES	EUKARYOTES
ORGANISMS REPRESENTED	BACTERIA AND CYANOBACTERIA	PROTISTS, FUNGI, PLANTS AND ANIMALS
CELL SIZE	SMALL, GENERALLY 1 TO 10 MICROMETERS	LARGE, GENERALLY 10 TO 100 MICROMETERS
METABOLISM AND PHOTOSYNTHESIS	ANAEROBIC OR AEROBIC	AEROBIC
MOTILITY	NONMOTILE OR WITH FLAGELLA MADE OF THE PROTEIN FLAGELLIN	USUALLY MOTILE, CILIA OR FLAGELLA CONSTRUCTED OF MICROTUBULES
CELL WALLS	OF CHARACTERISTIC SUGARS AND PEPTIDES	OF CELLULOSE OR CHITIN, BUT LACKING IN ANIMALS
ORGANELLES	NO MEMBRANE-BOUNDED ORGANELLES	MITOCHONDRIA AND CHLOROPLASTS
GENETIC ORGANIZATION	LOOP OF DNA IN CYTOPLASM	DNA ORGANIZED IN CHROMOSOMES AND BOUNDED BY NUCLEAR MEMBRANE
REPRODUCTION	BY BINARY FISSION	BY MITOSIS OR MEIOSIS
CELLULAR ORGANIZATION	MAINLY UNICELLULAR	MAINLY MULTICELLULAR, WITH DIFFERENTIATION OF CELLS

GREATEST DIVISION among organisms is the one separating cells with nuclei (eukaryotes) from those without nuclei (prokaryotes). The only prokaryotes are bacteria and cyanobacteria, and here they are represented by the bacterium *Escherichia coli* (*top left*). All other organisms are eukaryotes, including higher plants and animals, fungi and protists such as *Euglena* (*top right*). Eukaryotic cells are by far the more complex ones, and some of the organelles they contain, such as mitochondria and chloroplasts, may be derived from prokaryotes that established a symbiotic relationship with the host cell. Prokaryotes vary widely in their tolerance of or requirement for free oxygen, and they are thought to have evolved during a period of fluctuating oxygen. All eukaryotes require oxygen for metabolism and for the synthesis of various substances, and they must have emerged after an atmosphere rich in oxygen became established.

ic cells takes place through the complicated process of mitosis, and most eukaryotes can also reproduce sexually through meiosis and the subsequent fusion of sex cells. (The "parasexual" reproduction of some prokaryotes differs markedly from advanced eukaryotic sexuality.) Eukaryotic cells are generally larger than prokaryotic ones, although the range of sizes overlaps, and almost all prokaryotes are unicellular organisms whereas the majority of eukaryotes are large, complex and many-celled. A mammalian animal, for example, can be made up of billions of cells, which are highly differentiated in both structure and function.

An intriguing feature of eukaryotic cells is that they have within them smaller membrane-bounded subunits, or organelles, the most notable being mitochondria and chloroplasts. Mitochondria are present in all eukaryotes, where they play a central role in the energy economy of the cell. Chloroplasts are present in some protists and in all green plants and are responsible for the photosynthetic activities of those organisms. It has been suggested that both mitochondria and chloroplasts may be evolutionary derivatives of what were once free-living microorganisms, an idea discussed in particular by Lynn Margulis of Boston University. The modern chloroplast, for example, may be derived from a cyanobacterium that was engulfed by another cell and that later established a symbiotic relationship with it. In support of this hypothesis it has been noted that both mitochondria and chloroplasts contain a small fragment of DNA whose organization is somewhat like that of prokaryotic DNA. In the past several years the testing of this hypothesis has generated a large body of

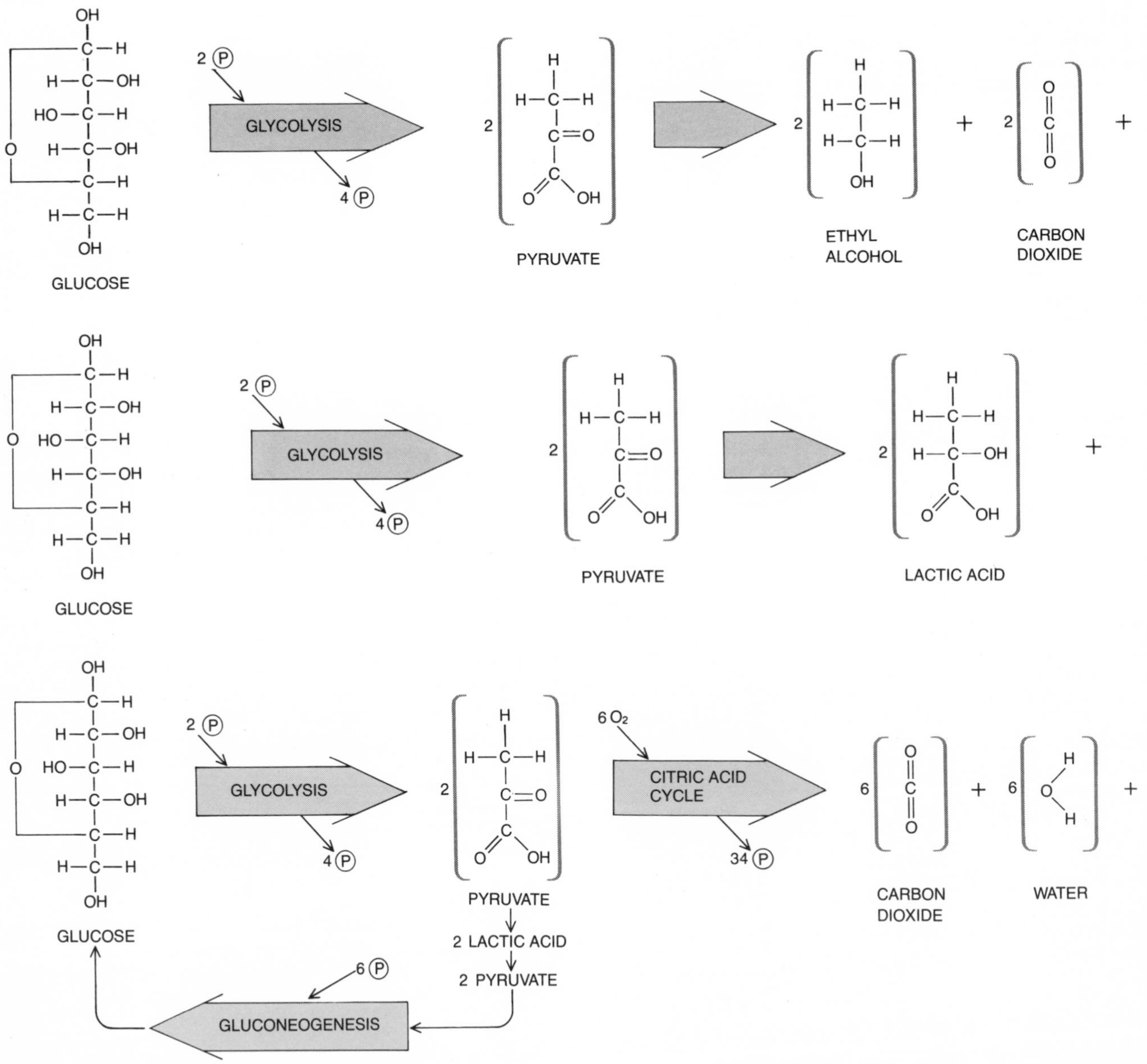

METABOLIC PATHWAYS by which cells extract energy from foodstuff apparently evolved in response to an increase in free oxygen. In all organisms the only usable energy derived from the breakdown of carbohydrates such as glucose is the fraction stored in high-energy phosphate bonds, denoted ~P; the rest is lost as heat. In anaerobic organisms (those that live without oxygen) glucose is broken down through fermentation: each molecule of glucose is split into two molecules of pyruvate, the process called glycolysis, with a net gain of two phosphate bonds. In bacterial fermentation the pyruvate is converted, in a step that provides no usable energy, into products such as lactic acid or ethyl alcohol and carbon dioxide, which are excreted a wastes. The metabolic system of aerobic organisms (those that requir oxygen) is respiration. It begins with glycolysis, but the pyruvate i treated not as a waste but as a substrate for a further series of re actions that make up the citric acid cycle. In these reactions pyru vate is decomposed one carbon atom at a time and combined wit oxygen, the ultimate products being carbon dioxide and water. Res piration releases far more energy than fermentation, and the propor tion of the energy recovered in useful form is also greater; as a re sult 36 phosphate bonds are formed instead of two. Respiratory metab

data on the comparative biochemistry of modern microorganisms, data that also provide clues to the evolution of life in the Precambrian.

One further difference between prokaryotes and eukaryotes is of particular importance in the study of their evolution: the extent to which the two types of organisms tolerate oxygen. Among the prokaryotes oxygen requirements are quite variable. Some bacteria cannot grow or reproduce in the presence of oxygen; they are classified as obligate anaerobes. Others can tolerate oxygen but can also survive in its absence; they are facultative anaerobes. There are also prokaryotes that grow best in the presence of oxygen but only at low concentrations, far below that of the present atmosphere. Finally, there are fully aerobic prokaryotes, forms that cannot survive without oxygen.

In contrast to this variety of adaptations the eukaryotes present a pattern of great consistency: with very few exceptions they have an absolute requirement for oxygen, and even the exceptions seem to be evolutionary derivatives of oxygen-dependent organisms. This observation leads to a simple hypothesis: the prokaryotes evolved during a period when environmental oxygen concentrations were changing, but by the time the eukaryotes arose the oxygen content was stable and relatively high.

One indication that eukaryotic cells have always been aerobic is provided by mitotic cell division, a process that can be considered a definitive characteristic of the group. Many eukaryotic cells can survive temporary deprivation of oxygen and can even carry on some metabolic functions; it appears that no cell, however, can undergo mitosis unless oxygen is available at least in low concentration.

The pathways of metabolism itself—the biochemical mechanisms by which an organism extracts energy from foodstuff—provide more detailed evidence. In eukaryotes the central metabolic process is respiration, which in overall terms can be described as the burning of the sugar glucose with oxygen to yield carbon dioxide, water and energy. Some prokaryotes (the aerobic or facultative ones) are also capable of respiration, but many derive their energy solely from the simpler process of fermentation. In bacterial fermentation glucose is not combined with oxygen (or with any other substance from outside the cell) but is simply broken down into smaller molecules. In both respiration and fermentation part of the energy released through the decomposition of glucose is captured in the form of high-energy phosphate bonds, usually in molecules of adenosine triphosphate (ATP). The rest of the energy is lost from the cell as heat.

Respiratory metabolism has two main components: a short series of chemical reactions, collectively called glycolysis, and a longer series called the citric acid cycle. In glycolysis a glucose molecule, with six carbon atoms, is broken down into two molecules of pyruvate, each having three carbon atoms. No oxygen is required for glycolysis, but on the other hand it releases only a little energy with a net gain of only two molecules of ATP.

The fuel for the citric acid cycle is the pyruvate formed by glycolysis. Through a series of enzyme-controlled reactions the carbon atoms of the pyruvate are oxidized and the oxidations are coupled to other reactions that result in the synthesis of ATP. For each two molecules of pyruvate (and hence for each molecule of glucose entering the sequence) 34 additional molecules of ATP are formed. The complete respiratory pathway is thus far more effective than glycolysis alone. In respiration the proportion of energy released that can be recovered in useful form (as ATP) is higher than it is in fermentation, about 38 percent instead of only some 30 percent, and in respiration the net energy yield to the cell is some 18 times greater. By breaking down the glucose to simple inorganic molecules (carbon dioxide and water) respiration liberates virtually all the biologically usable energy stored in the chemical bonds of the sugar.

The metabolism of the prokaryotes immediately suggests an evolutionary relationship between them and the eukaryotes: up to a point fermentation is indistinguishable from glycolysis. In bacterial fermentation a molecule of glucose is split into two molecules of pyruvate, with a net yield of two molecules of ATP. As in glycolysis, no oxygen is required for the process. In anaerobic prokaryotes, however, the metabolic pathway essentially ends at pyruvate. The only further reactions transform the pyruvate into such compounds as lactic acid, ethyl alcohol or carbon dioxide, which are excreted by the cell as wastes.

The similarity of fermentation in prokaryotes to glycolysis in eukaryotes seems too close to be a coincidence, and the assumption of an evolutionary relationship between the two groups provides a ready explanation. It seems likely that anaerobic fermentation became established as an energy-yielding process early in the history of life. When atmospheric oxygen became available for metabolism, it offered the potential for extracting 18 times as much useful energy from carbohydrate: a net yield of 36 molecules of ATP instead of only two molecules. The oxygen-dependent reactions did not, however, simply replace the anaerobic ones; they were appended to the existing anaerobic pathway.

Further evidence for this proposed evolutionary sequence can be found in the behavior of some eukaryotic cells under conditions of oxygen deprivation. In mammalian muscle cells, for example, prolonged exertion can demand more oxygen than the lungs and the blood can supply. The citric acid cycle is then disabled, but the cells continue to function, albeit at reduced efficiency, through glycolysis alone. Under such conditions of oxygen debt pyruvate is not consumed in the cell, but in the liver it can be converted back into glucose (at a cost in energy of six ATP molecules). Significantly the pyruvate itself is not transported to the liver but instead is

TOTAL ENERGY RELEASED Ⓟ PLUS HEAT (KILOCALORIES)	AVAILABLE ENERGY Ⓟ (KILOCALORIES)	CALCULATED EFFICIENCY (PERCENT)
57	14.6	26
47	14.6	31
686	262.8	38

olism could have evolved, however, only when free oxygen became readily available; it appears to have developed simply by appending the citric acid cycle to the glycolytic pathway. When aerobic cells are deprived of oxygen, many revert to fermentative metabolism, converting pyruvate into lactic acid. In vertebrates the lactic acid from muscle cells is exported to the liver, where it is returned to the form of pyruvate and converted into glucose.

converted into lactic acid, which in the liver must then be returned to the form of pyruvate. This use of lactic acid may represent a vestige of an earlier, bacterial pathway that under aerobic conditions has been suppressed. Indeed, the oxygen-starved muscle cell seems to revert to a more primitive, entirely anaerobic form of metabolism.

The development of an oxygen-dependent biochemistry can also be traced through a consideration of reaction sequences in the synthesis of various biological molecules. Once again stages in the synthetic pathway that emerged early in the Precambrian can be expected to proceed in the absence of oxygen. Reaction steps nearer the end product of the pathway, which were presumably added at a later age, might with

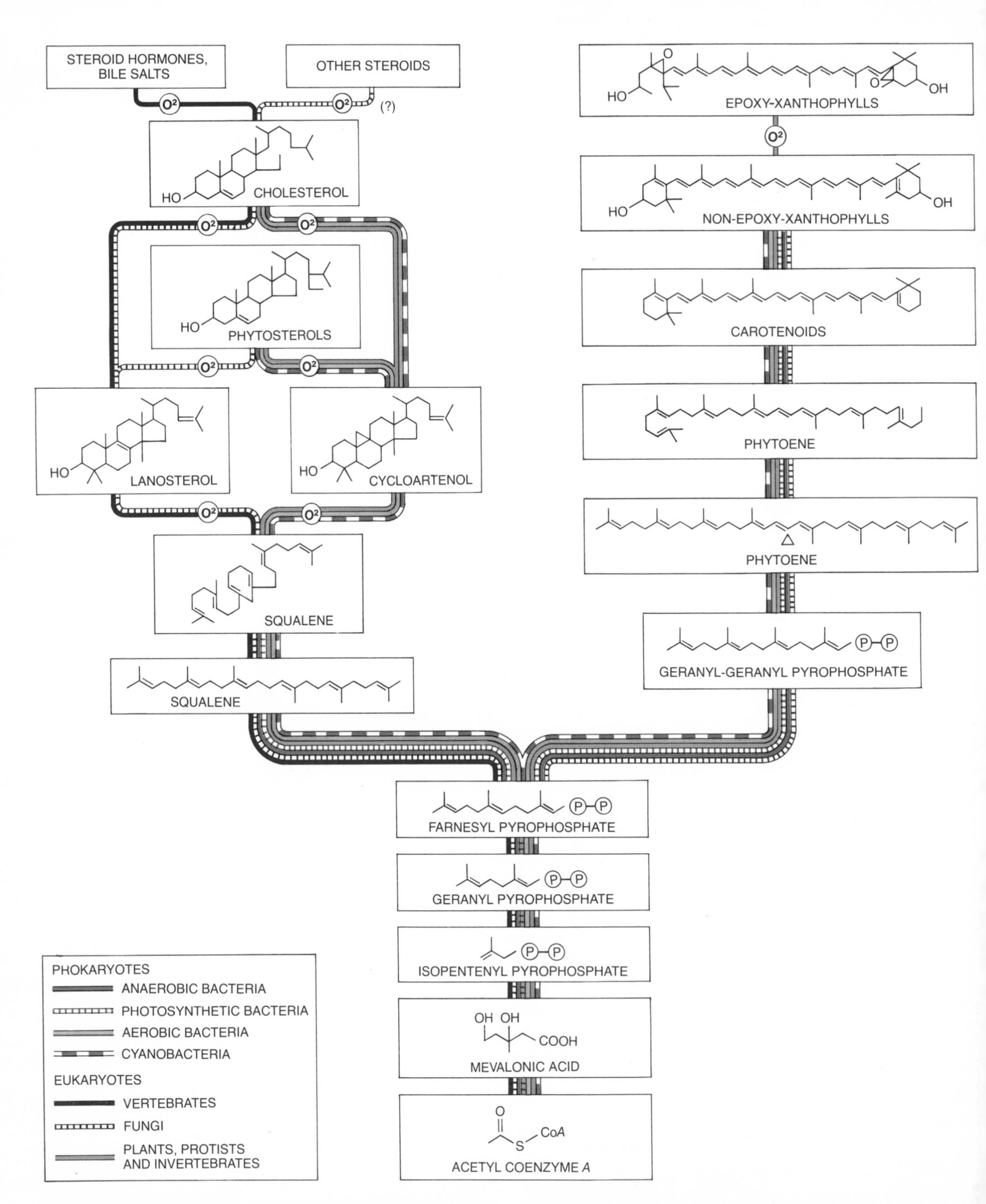

SYNTHESIS OF STEROLS and of related compounds such as the carotenoid pigments of plants requires molecular oxygen (O_2) only for steps near the end of the reaction sequence. Oxygen-dependent steps can be carried out only by aerobic organisms that evolved comparatively late in history of Precambrian life. Organic molecules are in schematic form with most carbon and hydrogen atoms omitted.

increasing frequency require oxygen. The distribution of the oxygen-demanding steps among various kinds of organisms could also have evolutionary significance. If only one pathway has evolved for the synthesis of a class of biochemical substances, then primitive forms of life might be expected to exhibit only the initial, anaerobic steps. Organisms that arose later might exhibit progressively longer, oxygen-dependent synthetic sequences.

In aerobic organisms it might seem at first that virtually all biochemical syntheses require oxygen; eukaryotic cells exhibit relatively little synthetic activity under anoxic conditions. For the most part, however, the oxygen requirement of such syntheses is simply for metabolism: the construction of biological molecules demands energy in the form

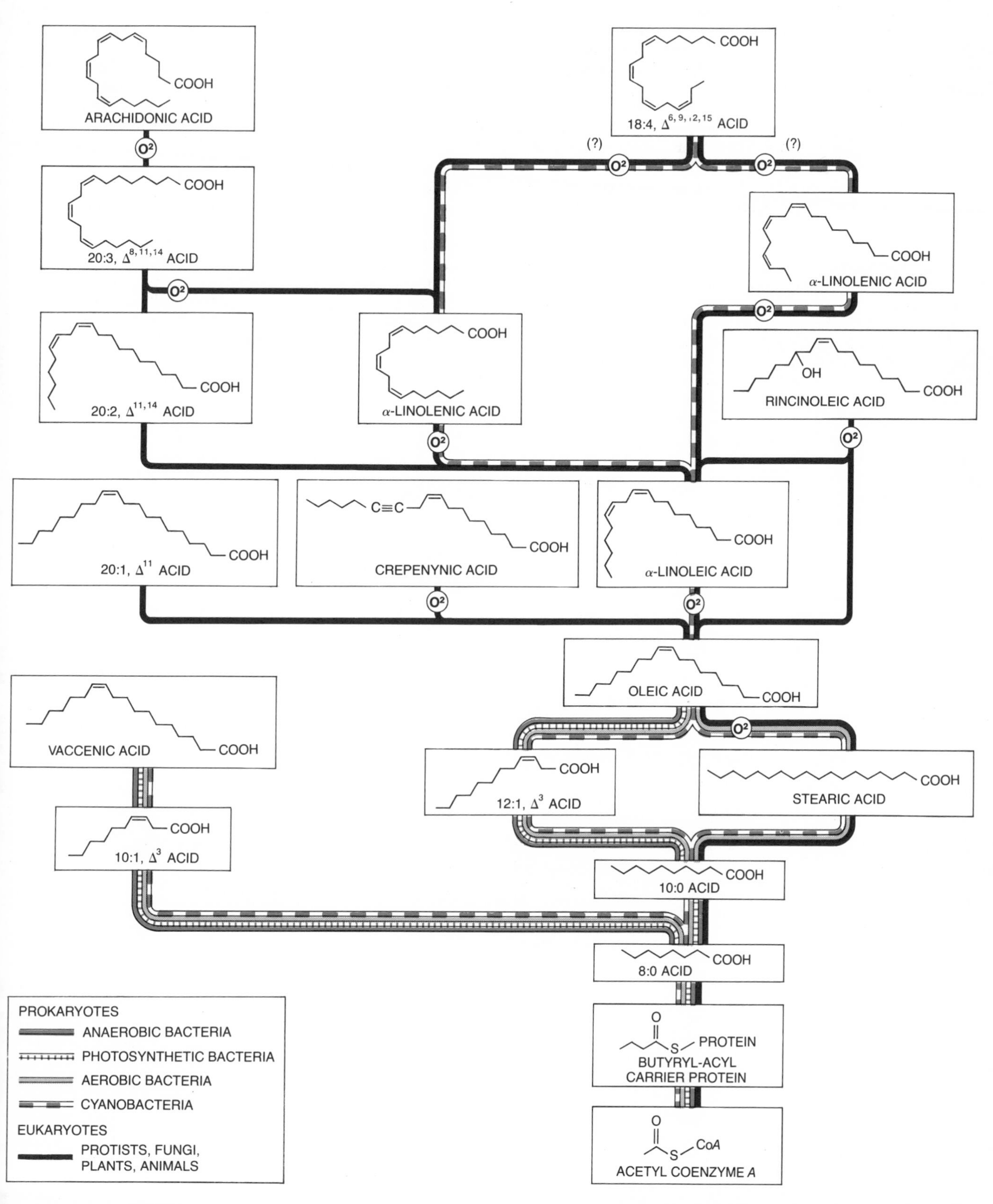

FATTY-ACID SYNTHESIS also follows a pattern suggesting the late addition of oxygen-dependent steps. Most prokaryotes can make mono-unsaturated fatty acids (those having one double bond) by inserting double bond during elongation of molecule. Eukaryotes and some prokaryotes first make a fully saturated molecule, stearic acid, then introduce double bonds by the process of oxidative desaturation.

of ATP, and most of the ATP is supplied through the oxygen-dependent citric acid cycle. If ATP is made available from some other source, many synthetic pathways can proceed unimpaired.

Some syntheses, however, have an intrinsic requirement for oxygen, quite apart from metabolic demands. Molecular oxygen is needed, for example, in the synthesis of bile pigments in vertebrates, of chlorophyll *a* in higher plants and of the amino acids hydroxyproline and, in animals, tyrosine. The oxygen dependence of two synthetic pathways in particular has been determined in detail. One of these pathways controls the manufacture of a class of compounds that includes the sterols and the carotenoids and the other is concerned with the synthesis of fatty acids.

Sterols, such as cholesterol and the steroid hormones, are flat, platelike molecules derived from the compound squalene, which has 30 carbon atoms. Carotenoids are derived from the 40-carbon compound phytoene; they are pigments, such as carotene, the orange-yellow compound in carrots, and they are found in virtually all photosynthetic organisms. A common starting point for the synthesis of both groups of compounds is isoprene, a five-carbon molecule that is also the repeating unit in synthetic rubber. In the biological synthesis two isoprene subunits are joined head to tail; then a third isoprene is added to form a 15-carbon polymer, farnesyl pyrophosphate. At this point there is a fork in the pathway. In one continuation of the synthesis two farnesyl chains are joined to form squalene, the 30-carbon precursor of the sterols. In the other continuation a fourth isoprene subunit is added, and only then are two of the chains joined. The product in this case is phytoene, the 40-carbon precursor of the carotenoids and of other pigments derived from them, such as the xanthophylls.

Up to this step in the synthetic pathway none of the reactions requires the participation of molecular oxygen. The next step in the synthesis of sterols, however, is the conversion of the linear squalene molecule to a 30-carbon ring, and this transformation does require oxygen; so do most of the subsequent steps in sterol synthesis. On the other branch of the pathway there are a few more anaerobic reactions, and indeed carotenoids can be made from phytoene without oxygen. Several further modifications of the carotenoids, however, such as the production of the pigments called epoxy-xanthophylls, are oxygen dependent.

Two observations about the evolution of these biosynthetic pathways are appropriate. Even in groups of organisms that have long been aerobic the first steps in the synthesis are independent of the oxygen supply; molecular oxygen enters the reaction sequence only at later stages. In a similar way the most primitive living organisms, the anaerobic bacteria, are capable only of the first segments of the pathway, the anaerobic segments. The more complex aerobic bacteria and the photosynthetic cyanobacteria have longer synthetic pathways, including some steps that require oxygen. Advanced eukaryotes, such as vertebrate animals and higher plants, have long, branched synthetic pathways, with many steps in which molecular oxygen is required.

A similar pattern can be discerned in the synthesis of fatty acids and their derivatives. The fatty acids are straight carbon-chain compounds that have a carboxyl group (COOH) at one end. A fatty acid is said to be saturated if there are no double bonds between carbon atoms in the chain; it is saturated with hydrogen, which fills all the available bonding positions. An unsaturated fatty acid has a double bond between two carbon atoms or it may have several such double bonds; for each double bond two hydrogen atoms must be removed from the molecule.

In the synthesis of fatty acids the molecule grows by the repeated addition of units two carbon atoms long. The first few steps in the synthesis are identical in all organisms, and they yield fully saturated fatty acids. The first branch in the pathway comes when the developing chain is eight carbons long. At that point many prokaryotes can introduce a double bond, which eukaryotes cannot. There is a second branch at the next step when the saturated chain is 10 carbons long; a double bond can similarly be introduced at that point by many prokaryotes but not by eukaryotes. No matter which branch is followed, elongation of the chain ends at 18 carbons. At that point the fatty acids produced by many prokaryotes contain a double bond, but in eukaryotes the product is always the fully saturated molecule, stearic acid. None of the steps in this sequence, whether in prokaryotes or eukaryotes, requires molecular oxygen.

If no subsequent transformations of fatty acids were possible, eukaryotic cells would be incapable of synthesizing any but the fully saturated forms. Actually extensive modifications can be accomplished through the process of oxidative desaturation, in which double bonds are formed by removing two hydrogen atoms and combining them with oxygen to form water. Oxidative desaturation can take place only in the presence of molecular oxygen (O_2). Through this mechanism cyanobacteria make unsaturated fatty acids with two, three and four double bonds, and eukaryotes form

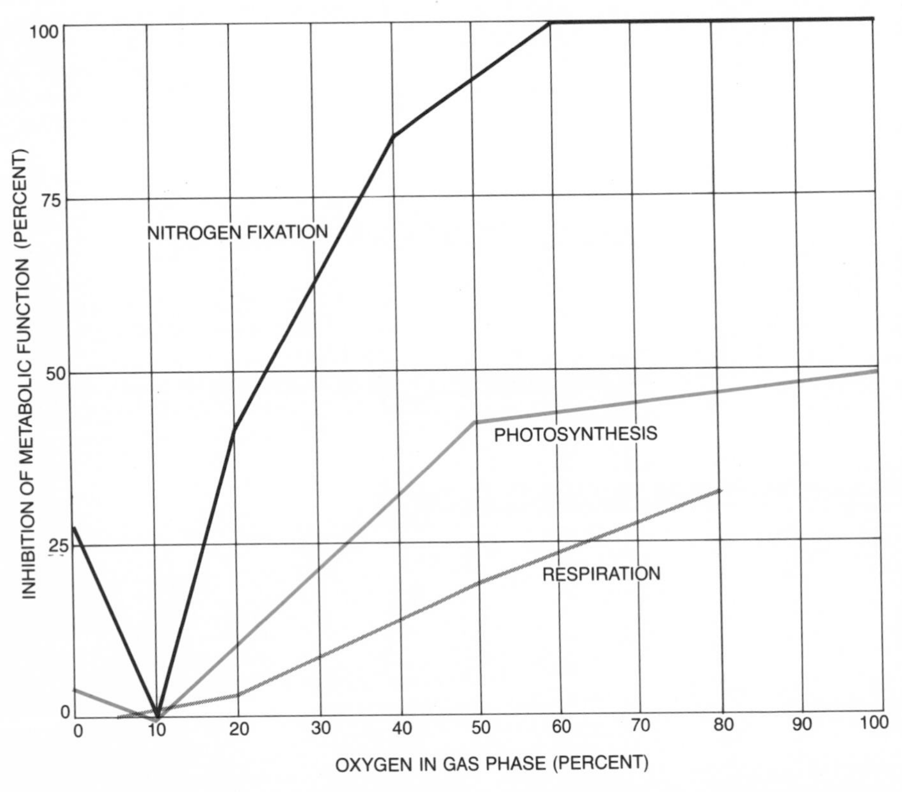

OXYGEN INHIBITION of metabolic functions in cyanobacteria suggests that these aerobic prokaryotes are adapted to an optimum oxygen concentration of about 10 percent, or roughly half the oxygen concentration of the earth's present atmosphere. Nitrogen fixation is completely halted by high oxygen levels, but even respiration, which requires oxygen, can be partially inhibited. Data are for the heterocyst-forming cyanobacterium *Anabaena flos-aquae.*

polyunsaturated fatty acids (with multiple double bonds).

As in the sterol-carotenoid synthesis, an analysis of the fatty-acid pathway argues for a pattern of biochemical evolution in which the increasing availability of atmospheric oxygen played a central role. The first steps in the synthetic sequence are common to all organisms capable of making fatty acids, and in the most primitive organisms those are the only steps. Hence the reactions that come first in the biochemical sequence apparently also developed early in the history of life; these first steps are all anaerobic. Organisms that presumably emerged somewhat later (such as aerobic bacteria and cyanobacteria) have longer pathways, including a few steps of oxidative desaturation. In advanced eukaryotes a substantial proportion of the steps are oxygen dependent.

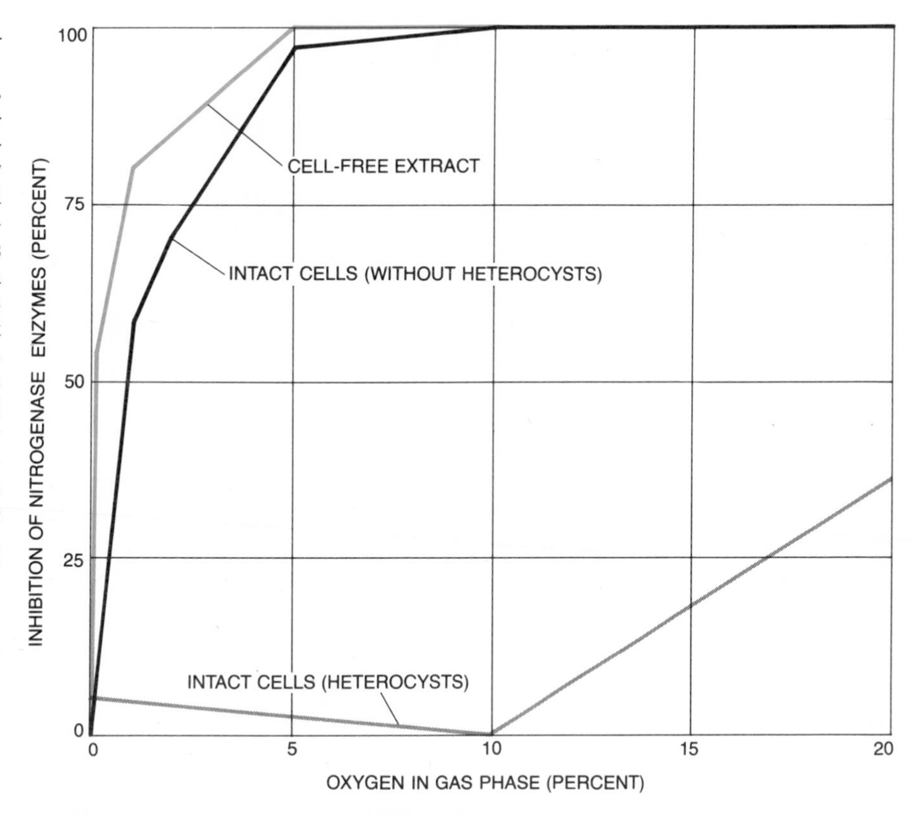

INHIBITION OF NITROGEN FIXATION in the presence of oxygen is caused by the deactivation of the nitrogenase enzymes. In cell-free extracts of the cyanobacterium *Plectonema boryanum* the nitrogenase enzymes are inhibited by even minute quantities of oxygen, and the intact cells of this species, which does not form heterocysts, offer little protection against the inhibition; such organisms can fix nitrogen only in an anoxic habitat. The thick cell walls and other special features of heterocyst cells, such as those formed by *Nostoc muscorum*, allow fixation to continue in a fully aerobic environment. Data suggest that the capability for nitrogen fixation evolved before significant quantities of oxygen had accumulated in the atmosphere.

Comparisons of the metabolism and biochemistry of prokaryotes and eukaryotes thus provide strong evidence that the latter group arose only after a substantial quantity of oxygen had accumulated in the atmosphere. Hence it is of interest to ask when eukaryotic cells first appeared. It seems apparent that an oxygen-rich atmosphere cannot have developed later than this signal evolutionary event.

The primary means of assigning a date to the origin of the eukaryotes is through the fossil record. Because this field of study is so new, however, the available information is scanty and often difficult to interpret. It is rarely a straightforward task to identify a microscopic, single-cell organism as being eukaryotic merely from an examination of its fossilized remains. And even when a fossil has been identified as unequivocally eukaryotic the available radioactive-isotope methods of dating can rarely assign it a precise age. At best such methods have an accuracy of only about plus or minus 5 percent. What is more, the age determinations are generally carried out on rocks that were once molten, such as volcanic lavas, whereas the fossils are found in sedimentary deposits. Consequently the stratum of the fossil itself usually cannot be dated; it is merely assigned an age somewhere between the ages of the nearest underlying and overlying datable rock units.

In spite of these difficulties there is now substantial evidence for the existence of eukaryotic fossils in rocks hundreds of millions of years older than the earliest Phanerozoic strata. The evidence is of two kinds: microfossils that display a morphological or organizational complexity judged to be of eukaryotic character, and the presence of fossil cells whose size is typical only of eukaryotes.

The evidence from relatively complex microscopic fossils includes the following: (1) branched filaments, made up of cells with distinct cross walls and resembling modern fungi or green algae, from the Olkhin formation of Siberia, a deposit thought to be about 725 million years old (but with a known age of between 680 and 800 million years); (2) complex, flask-shaped microfossils from the Kwagunt formation in the eastern Grand Canyon, thought to be about 800 (or 650 to 1,150) million years old; (3) fossils of unicellular algae containing intracellular membranes and small, dense bodies that may represent preserved organelles, from the Bitter Springs formation of central Australia, dated at approximately 850 (or 740 to 950) million years; (4) a group of four sporelike cells in a tetrahedral configuration that may have been produced by mitosis or possibly meiosis, also from Bitter Springs rocks; (5) spiny cells or algal cysts several hundred micrometers in diameter and with unquestionable affinities to eukaryotic organisms, from Siberian shales that are reportedly 950 (or 750 to 1,050) million years old; (6) highly branched filaments of large diameter and with rare cross walls, similar in some respects to certain green or golden-green eukaryotic algae, from the Beck Spring dolomite of southeastern California (1,300, or 1,200 to 1,400 million years old) and from the Skillogalee dolomite of South Australia (850, or 740 to 867 million years old); (7) spheroidal microfossils described as exhibiting two-layered walls and having "medial splits" on their surface, and which may represent an encystment stage of a eukaryotic alga, from shales 1,400 (or 1,280 to 1,450) million years old in the McMinn formation of northern Australia; (8) a tetrahedral group of four small cells, resembling spores produced by mitotic cell division of some green algae, from the Amelia dolomite of northern Australia, approaching 1,500 (or 1,390 to 1,575) million years in age; (9) unicellular fossils that appear to be exceptionally well preserved and that are reported to contain small membrane-bounded structures that could be remnants of organelles, from the Bungle-Bungle dolomite in the same region as the Amelia dolomite and of approximately the same age.

Thus the earliest of these eukaryote-like fossils are probably somewhat less than 1,500 million years old. Numerous types of microfossils have been discovered in older sediments, but none of them seems to be a strong candidate for

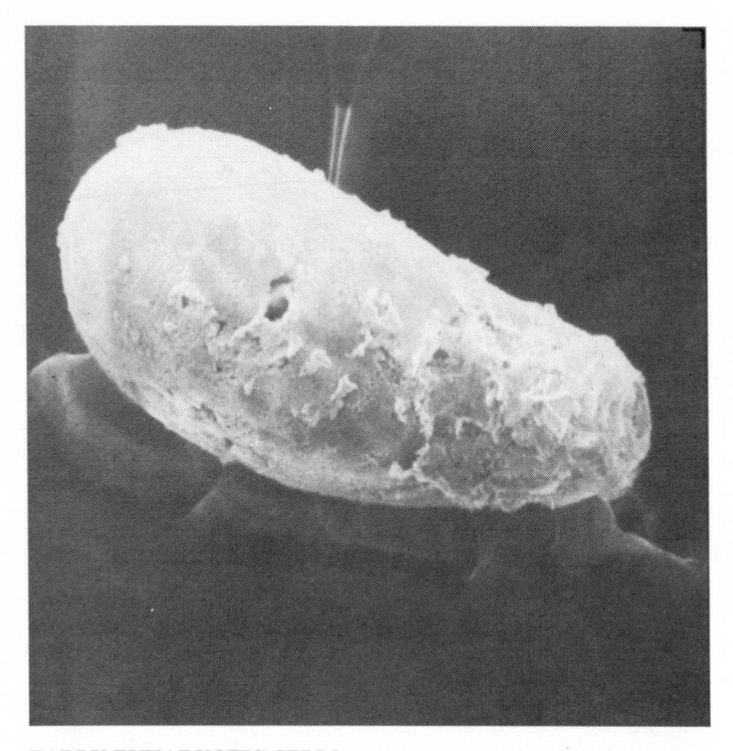

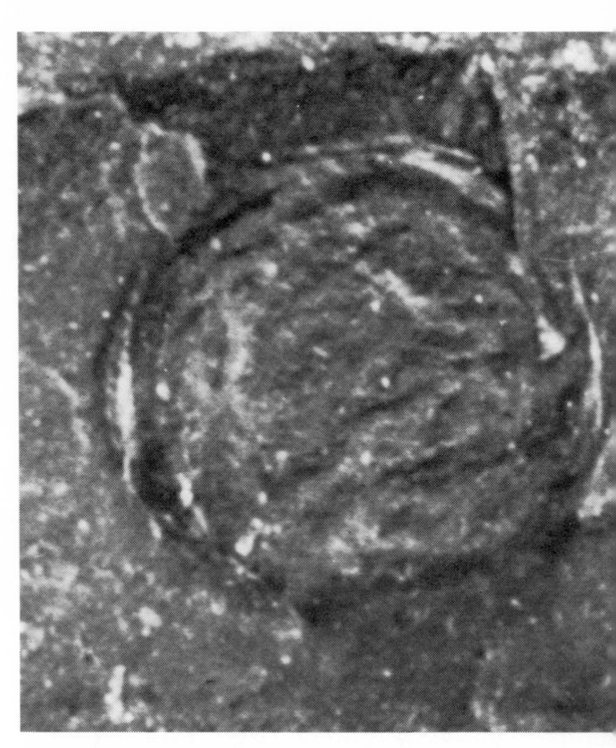

EARLY EUKARYOTIC CELLS may be represented among Precambrian microfossils. The gourd-shaped cell at the left, from shales in the Grand Canyon thought to be 800 million years old, is morphologically more complex than any known prokaryote; it is also larger, about 100 micrometers long. The second cell from the left is some two millimeters in diameter and hence is more than 30 times the size of

identification as eukaryotic. For example, the well-studied Canadian fossils of the Gunflint and Belcher Island iron formations, which are about two billion years old, have been interpreted as exclusively prokaryotic.

The testimony of these as yet rare and unusual specimens can be checked through statistical studies of the sizes of known Precambrian microfossils. The size ranges of prokaryotes and eukaryotes overlap, so that a particular fossil cannot always be classified unambiguously on the basis of size alone; by cataloguing the measured sizes in a large sample of fossils, however, it may be possible to determine whether or not eukaryotic cells are present. Among modern species of spheroidal cyanobacteria about 60 percent are very small, less than five micrometers in diameter; of the remaining species only a few are larger than 20 micrometers and none is larger than 60 micrometers. Unicellular eukaryotes, such as green or red algae, can be much larger. Typically they fall in the range between five and 60 micrometers, but several percent of living species are larger than 60 micrometers and a few are larger than 1,000 micrometers (one millimeter).

Systematic size measurements have been made on some 8,000 fossil cells from 18 widely dispersed Precambrian deposits. On the basis of those data certain tentative conclusions can be drawn. Cells larger than 100 micrometers, and hence of distinctly eukaryotic dimensions, are unknown in rocks older than about 1,450 million years. Virtually all the unicellular fossils from rocks of that age, whether they grew in shallow-water stromatolites or were deposited in offshore shales, are of prokaryotic size.

Cells larger than modern prokaryotes (greater than 60 micrometers in diameter) first become abundant in rocks about 1,400 million years old. Algae of this type were apparently free-floating rather than mat-forming species, and they are therefore particularly common in shales, sediments deposited in deeper water. Such eukaryote-size fossils have been known for several years from shales of this age in China and in the U.S.S.R. Recently cells more than 100 micrometers in diameter have also been discovered in the Newland limestone of Montana, and cells more than 600 micrometers in size (10 times the size of the largest spheroidal prokaryote) have been found in the McMinn formation of Australia; the age of both of these fossil-bearing deposits is about 1,400 million years.

In somewhat younger Precambrian sediments there are still larger cells, fossils greater than one millimeter in diameter (with some as large as eight millimeters). They were first described in 1899 by Walcott, who discovered them in rocks from the Grand Canyon. They have since been found in nearly a dozen other rock units throughout the world. The oldest seem to be those from Utah and from Siberia, each about 950 million years old, and those from northern India, which could be even older (from 910 to 1,150 million years old).

Studies of both the morphology and the size of unicellular fossils therefore suggest that there is a break in the fossil record between 1,400 and 1,500 million years ago. Below this horizon cells with eukaryotelike traits are rare or absent; above it they become increasingly common. Moreover, the data suggest that the diversification of the eukaryotes began shortly after the cell type first appeared, apparently within the next few hundred million years. By a billion years ago there had been substantial increases in cell size, in morphological complexity and in the diversity of species. All these indicators also suggest, of course, that oxygen-dependent metabolism, which is highly developed even in the most primitive eukaryotes, had already become established by about 1.5 billion years ago.

The prokaryotes that must have held exclusive sway over the earth before the development of eukaryotic cells were less diverse in form, but they were probably more varied in metabolism and biochemistry than their eukaryotic descendants. Like modern prokaryotes, the ancient species presumably varied over a broad range in their tolerance of oxygen, all the way from complete intolerance to absolute need. In this regard one group of prokaryotes, the cyanobacteria, are of particular interest in that they were largely responsible for the development of an oxygen-rich atmosphere.

Like higher plants, cyanobacteria carry out aerobic photosynthesis, a process

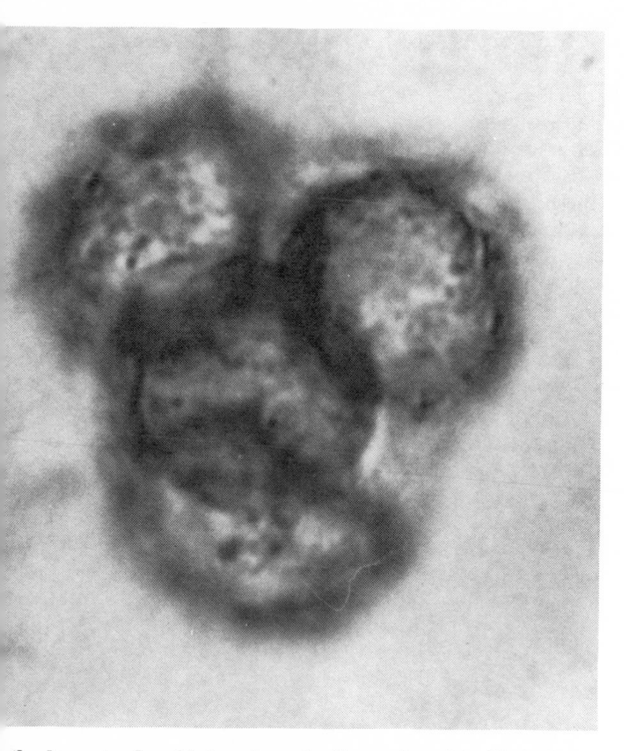

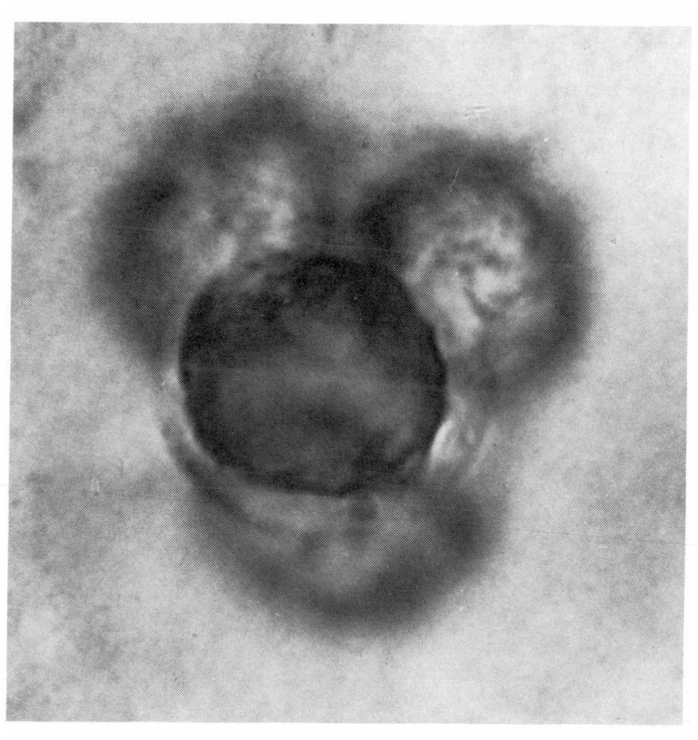

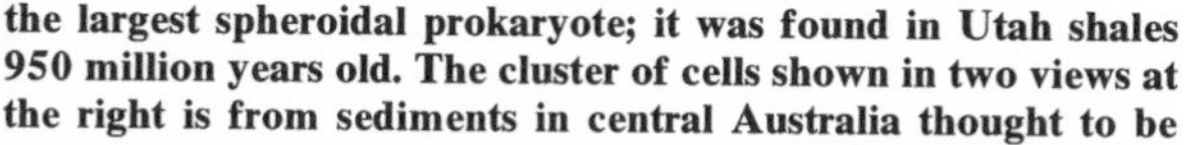

the largest spheroidal prokaryote; it was found in Utah shales 950 million years old. The cluster of cells shown in two views at the right is from sediments in central Australia thought to be 850 million years old. The cells are only 10 micrometers across, but their tetrahedral arrangement suggests they formed as a result of mitosis or possibly meiosis, mechanisms of cell division known only in eukaryotes.

that in overall effect (although not in mechanism) is the reverse of respiration. The energy of sunlight is employed to make carbohydrates from water and carbon dioxide, and molecular oxygen is released as a by-product. The cyanobacteria can tolerate the oxygen they produce and can make use of it both metabolically (in aerobic respiration) and in synthetic pathways that seem to be oxygen dependent (as in the synthesis of chlorophyll *a*). Nevertheless, the biochemistry of the cyanobacteria differs from that of green, eukaryotic plants and suggests that the group originated during a time of fluctuating oxygen concentration. For example, although many cyanobacteria can make unsaturated fatty acids by oxidative desaturation, some of them can also employ the anaerobic mechanism of adding a double bond during the elongation of the chain. In a similar manner oxygen-dependent syntheses of certain sterols can be carried out by some cyanobacteria, but the amounts of the sterols made in his way are minuscule compared with he amounts typical of eukaryotes. In other cyanobacteria those sterols are not found at all, the biosynthetic pathway being terminated after the last anaerobic step: the formation of squalene. Hence in their biochemistry the cyanobacteria seem to occupy a middle ground between the anaerobes and the ukaryotes.

In metabolism too the cyanobacteria occupy an intermediate position. They flourish today in fully oxygenated environments, but physiological experiments indicate that for many species optimum growth is obtained at an oxygen concentration of about 10 percent, which is only half that of the present atmosphere. Both photosynthesis and respiration are increasingly inhibited when the oxygen concentration exceeds that optimum level. It has recently been discovered that some cyanobacteria can switch the cellular machinery of aerobic metabolism on and off according to the availability of oxygen. Under anoxic conditions these species not only halt respiration but also adopt an anaerobic mode of photosynthesis, employing hydrogen sulfide (H_2S) instead of water and releasing sulfur instead of oxygen. This capability for anaerobic metabolism is probably a relic of an earlier stage in the evolutionary development of the group.

Another activity of some cyanobacteria that seems to reflect an earlier adaptation to anoxic conditions is nitrogen fixation. Nitrogen is an essential element of life, but it is biologically useful only in "fixed" form, for example combined with hydrogen in ammonia (NH_3). Only prokaryotes are capable of fixing nitrogen (although they often do so in symbiotic relationships with higher plants). The crucial complex of enzymes for fixation, the nitrogenases, is highly sensitive to oxygen. In cell-free extracts nitrogenases are partially inhibited by as little as .1 percent of free oxygen, and they are irreversibly inactivated in minutes by exposure to oxygen concentrations of only about 5 percent.

Such a complex of enzymes could have originated only under anoxic conditions, and it can operate today only if it is protected from exposure to the atmosphere. Many nitrogen-fixing bacteria provide that protection simply by adopting an anaerobic habitat, but among the cyanobacteria a different strategy has developed: the nitrogenase enzymes are protected in specialized cells, called heterocysts, whose internal milieu is anoxic. The heterocysts lack certain pigments essential for photosynthesis, and so they generate no oxygen of their own. They have thick cell walls and are surrounded by a mucilaginous envelope that retards the diffusion of oxygen into the cell. Finally, they are equipped with respiratory enzymes that quickly consume any uncombined oxygen that may leak in.

Because of the thick cell walls heterocysts should be comparatively easy to recognize in fossil material. Indeed, possible heterocysts have been reported from several Precambrian rock units, the oldest being about 2.2 billion years in age. If these cells are indeed heterocysts, they may be taken as a sign that free oxygen was present by then, at least in small concentrations.

Nitrogen fixation has a high cost in energy, and the capability for it would therefore seem to confer a selective advantage only when fixed nitrogen is a scarce resource. Today the main sources of fixed nitrogen are biological and industrial, but biologically usable nitrate (NO_3^-) is formed by the reaction of atmospheric nitrogen and oxygen. In the anoxic atmosphere of the early Precambrian the latter mechanism would obviously have been impossible. The lack of atmospheric oxygen would also have indirectly reduced the concentration of ammonia to very low levels. Ammonia is dissociated into nitrogen and hydrogen by ultraviolet radiation, most of which is filtered out today by a layer of ozone (O_3) high in the atmosphere; without free oxygen there would have been little ozone, and without this protective shield atmospheric ammonia would have been quickly destroyed.

It is likely that the capability for nitrogen fixation developed early in the Precambrian among primitive prokaryotic organisms and in an environment where fixed nitrogen was in short supply. The vulnerability of the nitrogenase enzymes to oxidation was of no consequence then, since the atmosphere had little oxygen. Later, as the photosynthetic activities of the cyanobacteria led to an increase in atmospheric oxygen, some nitrogen fixers adopted an anaerobic habitat and others developed heterocysts. By the time eukaryotes appeared, apparently more than half a billion years later, oxygen was abundant and fixed nitrogen (both NH_3 and NO_3^-) was probably less scarce, and so the eukaryotes never developed the enzymes needed for nitrogen fixation.

At present oxygen-releasing photosynthesis by green plants, cyanobacteria and some protists is responsible for the synthesis of most of the world's organic matter. It is not, however, the only mechanism of photosynthesis. The alternative systems are confined to a few groups of bacteria that on a global scale seem to be of minor importance today but that may have been far more significant in the geological past.

The several groups of photosynthetic bacteria differ from one another in their pigmentation, but they are alike in one important respect: unlike the photosynthesis of cyanobacteria and eukaryotes, all bacterial photosynthesis is a totally anaerobic process. Oxygen is not given off as a by-product of the reaction, and the photosynthesis cannot proceed in the presence of oxygen. Whereas oxygen appears to be required in green plants for the synthesis of chlorophyll *a*, oxygen inhibits the synthesis of bacteriochlorophylls.

The anaerobic nature of bacterial photosynthesis seems to present a paradox: photosynthetic organisms thrive where light is abundant, but such environments are also generally ones having high concentrations of oxygen, which poisons bacterial photosynthesis. These contradictory needs can be explained if it is assumed that anaerobic photosynthesis evolved among primitive bacteria early in the Precambrian, when the atmosphere was essentially anoxic. The photosynthesizers could thus have lived in matlike communities in shallow water and in full sunlight.

Somewhat later such bacteria gave rise to the first organisms capable of aerobic photosynthesis, the precursors of modern cyanobacteria. For the anaerobic photosynthetic bacteria the molecular oxygen released by this mutant strain was a toxin, and as a result the aerobic photosynthesizers were able to supplant the anaerobic ones in the upper portions of the mat communities. The anaerobic species became adapted to the lower parts of the mat, where there is less light but also a lower concentration of oxygen. Many photosynthetic bacteria occupy such habitats today.

Photosynthetic bacteria were surely not the first living organisms, but the history of life in the period that preceded their appearance is still obscure. What little information can be inferred about that early period, however, is consistent with the idea that the environment was then largely anoxic. One tentative line of evidence rests on the assumption that among organisms living today those that are simplest in structure and in biochemistry are probably the most closely related to the earliest forms of life. Those simplest organisms are bacteria of the clostridial and methanogenic types, and they are all obligate anaerobes.

There is even a basis for arguing that anoxic conditions must have prevailed during the time when life first emerged on the earth. The argument is based on the many laboratory experiments that have demonstrated the synthesis of organic compounds under conditions simulating those of the primitive planet. These syntheses are inhibited by even small concentrations of molecular oxygen. Hence it appears that life probably would not have developed at all if the early atmosphere had been oxygen-rich. It is also significant that the starting materials for such experiments often include hydrogen sulfide and carbon monoxide (CO), and that an intermediate in many of the reactions is hydrogen cyanide (HCN). All three compounds are poisonous gases, and it seems paradoxical that they should be forerunners of the earliest biochemistry. They are poisonous, however, only for aerobic forms of life; indeed, for many anaerobes hydrogen sulfide not only is harmless but also is an important metabolite.

It was argued above that oxygen must have been freely available by the time the first eukaryotic cells appeared, probably 1,400 to 1,500 million years ago. Hence the proliferation of cyanobacteria that released the oxygen must have taken place earlier in the Precambrian. How much earlier remains in question. The best available evidence bearing on the issue comes from the study of sedimentary minerals, some of which may have been influenced by the concentration of free oxygen at the time they were deposited. In recent years a number of workers have investigated this possibility, most notably Preston E. Cloud, Jr., of the University of California at Santa Barbara and the U.S. Geological Survey.

One mineral of significance in this argument is uraninite (UO_2), which is found in several deposits that were laid down in Precambrian streambeds. In the presence of oxygen, grains of uraninite are readily oxidized (to U_3O_8) and are thereby dissolved. David E. Grandstaff of Temple University has shown that streambed deposits of the mineral probably could not have accumulated if the concentration of atmospheric oxygen was greater than about 1 percent. Uraninite-bearing deposits of this type are found in sediments older than about two billion years but not in younger strata, suggesting that the transition in oxygen concentration may have come at about that time.

Another kind of mineral deposit, the iron-rich formations called red beds, exhibits the opposite temporal pattern: red beds are known in sedimentary sequences younger than about two billion years but not in older ones. The red beds are composed of particles coated with iron oxides (mostly the mineral hematite, Fe_2O_3), and many are thought to have formed by exposure to oxygen in the atmosphere rather than under water. It has been proposed that the oxygen may have been biologically generated. This hypothesis is consistent with several lines of evidence, but objections to it have also been raised. For example, most red beds are continental deposits rather than marine ones and are therefore susceptible to erosion; it is thus conceivable that red beds were formed earlier than two billion years ago as well as later but that the earlier beds have been destroyed. It is also possible that the oxygen in the red beds had a nonbiological origin; it may have come from the splitting of water by ultraviolet radiation. This has apparently happened on Mars to create a vast red bed across the surface of that planet, where there are only traces of free oxygen and there is no evidence of life.

Perhaps the most intriguing mineral evidence for the date of the oxygen transition comes from another kind of iron-rich deposit: the banded iron formation. These deposits include some tens of billions of tons of iron in the form of oxides embedded in a silica-rich matrix; they are the world's chief economic reserves of iron. A major fraction of them was deposited within a comparatively brief period of a few hundred million years beginning somewhat earlier than two billion years ago.

A transition in oxygen concentration could explain this major episode of iron sedimentation through the following hypothetical sequence of events. In a primitive, anoxic ocean, iron existed in the ferrous state (that is, with a valence of +2) and in that form was soluble in seawater. With the development of aerobic photosynthesis small concentrations of oxygen began diffusing into the upper portions of the ocean, where it reacted with the dissolved iron. The iron was thereby converted to the ferric form (with a valence of +3), and as a result hydrous ferric oxides were precipitated and accumulated with silica to form rusty layers on the ocean floor. As the process continued virtually all the dissolved iron in the ocean basins was precipitated: in a matter of a few hundred million years the world's oceans rusted.

As in the deposition of red beds, an inorganic origin could also be proposed for the oxygen in the banded iron formations; the oxygen in some formations laid down in the very early Precambrian might well have come from such a source. For the extensive iron formations of about two billion years ago, however, inorganic processes such as the photochemical splitting of water do not appear to be adequate; they probably could not have produced the necessary quantity of oxygen quickly enough to account for the enormous volume of iron ores deposited at about that time. Indeed, only one mechanism is known that could release oxygen at the required rate: aerobic photosynthesis, followed by the sedimentation and burial of the organic matter thereby produced. (The burial is a necessary condition, since aerobic decomposition of the organic remains would use up as much oxygen as had been generated.)

In relation to this hypothesis it is notable that fossil stromatolites first become abundant in sediments deposited about 2,300 million years ago, shortly before the major episode of iron-ore deposition. It is therefore possible that the first widespread appearance of stromatolites might mark the origin and the earliest diversification of oxygen-producing cyanobacteria. Even at that early date the cyanobacteria would probably have released oxygen at a high rate, but for several hundred million years the iron dissolved in the oceans would have served as a buffer for the oxygen concentration of the atmosphere, reacting with the gas and precipitating it as ferric oxides almost as quickly as it was generated. Only when the oceans had been swept free of unoxidized iron and similar materials would the concentration of oxygen in the atmosphere have begun to rise toward modern levels.

Although much remains uncertain, evidence from the fossil record, from modern biochemistry and from geology and mineralogy make possible a tentative outline for the history of Precam-

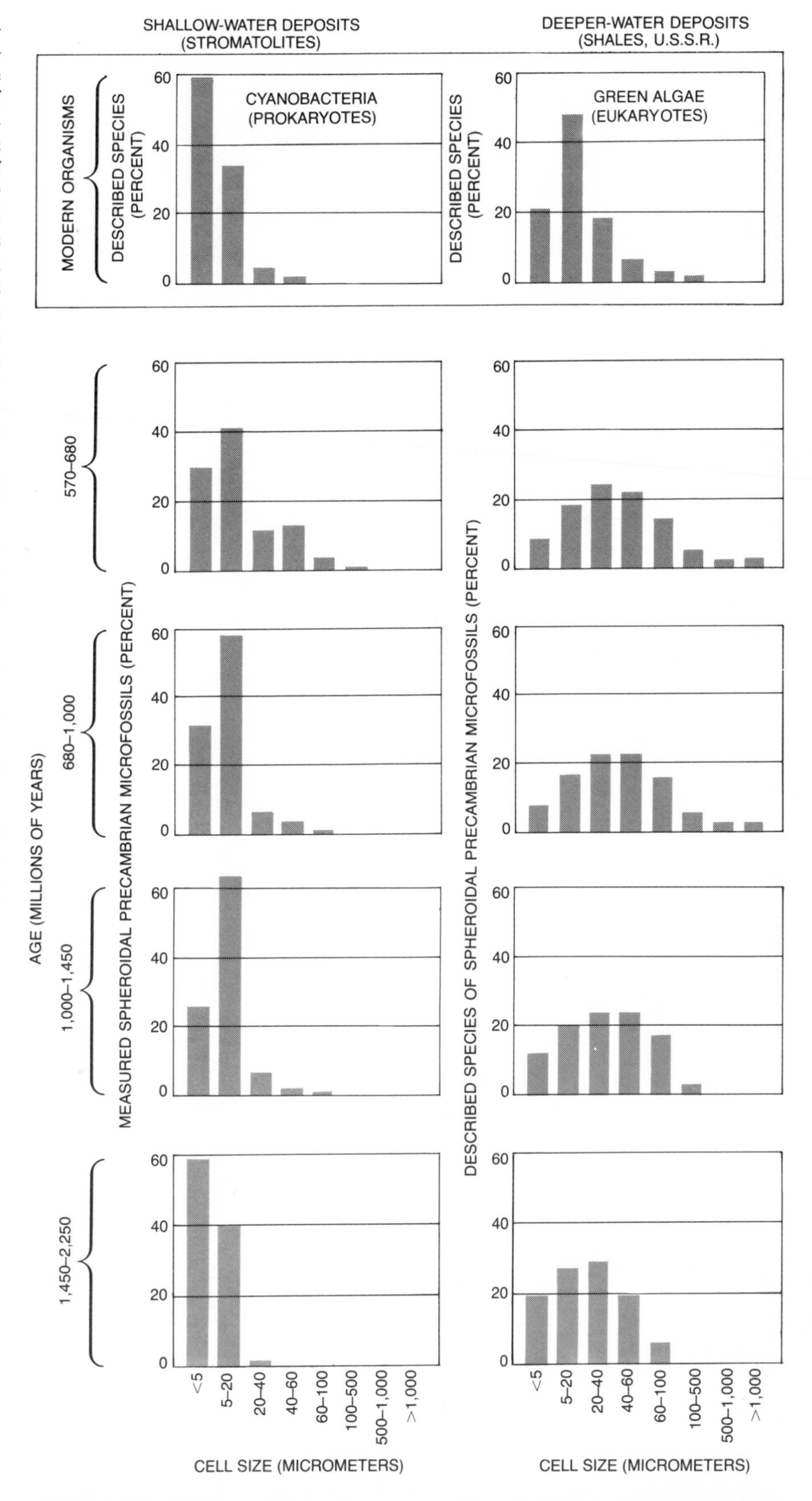

SIZE OF FOSSIL CELLS provides evidence on the origin of the eukaryotes. Spheroidal microfossils of various ages were measured and assigned to eight size categories; a similar procedure was followed with spheroidal members of two groups of modern microorganisms, the prokaryotic cyanobacteria and the eukaryotic green algae. The range of sizes for the modern species overlaps, but the largest cells are observed only among the eukaryotes. The oldest fossils examined have a distribution of sizes similar to that of prokaryotes, but Precambrian rock units younger than 1,450 million years include larger cells that are probably eukaryotic, and the proportion of larger cells increases in later periods. The larger cells tend to be more abundant in shales than in stromatolitic sediments. Because shales are deposited offshore that fact would be explained if early eukaryotes were predominantly free-floating rather than mat-forming.

brian life. The most primitive forms of life with recognizable affinities to modern organisms were presumably spheroidal prokaryotes, perhaps comparable to modern bacteria of the clostridial type. Initially at least they probably derived their energy from the fermentation of materials that were organic in nature but were of nonbiological origin. These materials were synthesized in the anoxic early atmosphere and were of the type that during the age of chemical evolution had led to the development of the first cells.

The first photosynthetic organisms apparently arose earlier than about three billion years ago. They were anaerobic prokaryotes, the precursors of modern photosynthetic bacteria. Most of them probably lived in matlike communities in shallow water, and they may have been responsible for building the earliest fossil stromatolites known, which are estimated to be about three billion years old.

The rise of aerobic photosynthesis in the mid-Precambrian introduced a change in the global environment that was to influence all subsequent evolution. The resulting increase in oxygen concentration probably led to the extinction of many anaerobic organisms, and others were forced to adopt mar-

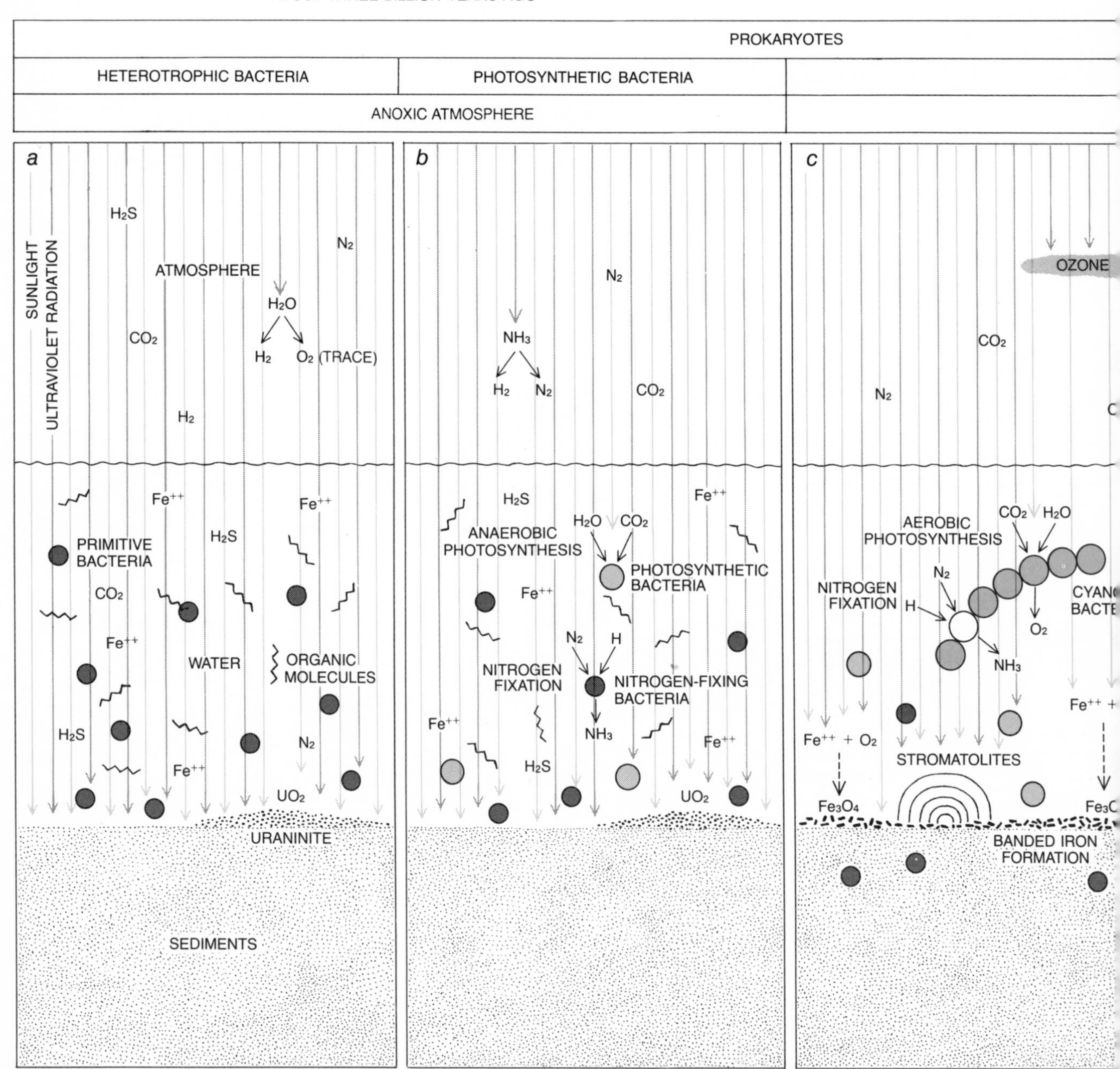

ORGANISM AND ENVIRONMENT evolved in counterpoint during the Precambrian. The first living cells (*a*) were presumably small, spheroidal anaerobes. Only traces of oxygen were present. They survived by fermenting organic molecules formed nonbiologically in the anoxic environment. The role of such ready-made nutrients was diminished, however, when the first photosynthetic organisms evolved (*b*). This earliest mode of photosynthesis was entirely anaerobic. Another early development was nitrogen fixation, required in part because ultraviolet radiation that could then freely penetrate the atmosphere would have quickly destroyed any ammonia (NH_3) present. A little more than two billion years ago (*c*) aerobic photosynthesis began in the precursors of modern cyanobacteria. Oxygen was generated by these stromatolite-building microorganisms, but for some 100 million years little of it accumulated in the atmosphere; instead it reacted with iron dissolved in the oceans, which was then precipitated to create massive banded iron formations. Only when the oceans had

ginal habitats, such as the lower reaches of bacterial mat communities. Nitrogen-fixing organisms also retreated to anaerobic habitats or developed heterocyst cells. With little competition for those regions having optimum light the cyanobacteria were able to spread rapidly and came to dominate virtually all accessible habitats. With the development of the citric acid cycle and its more efficient extraction of energy from foodstuff, the dominance of the biological community by aerobic organisms was confirmed. When the major episode of deposition of banded iron formations ended some 1,800 million years ago, the trend toward increasing oxygen concentration became irreversible.

By the time eukaryotic cells arose 1,500 to 1,400 million years ago a stable, oxygen-rich atmosphere had long prevailed. Adaptive strategies needed by earlier organisms to cope with fluctuations in the oxygen level were unnecessary for eukaryotes, which were from the start fully aerobic. The diversity of eukaryote cell types present by about a billion years ago suggests that some form of sexual reproduction may have evolved by then. Within the next 400 million years the rapid diversification of eukaryotic organisms had led to the emergence of multicellular forms of

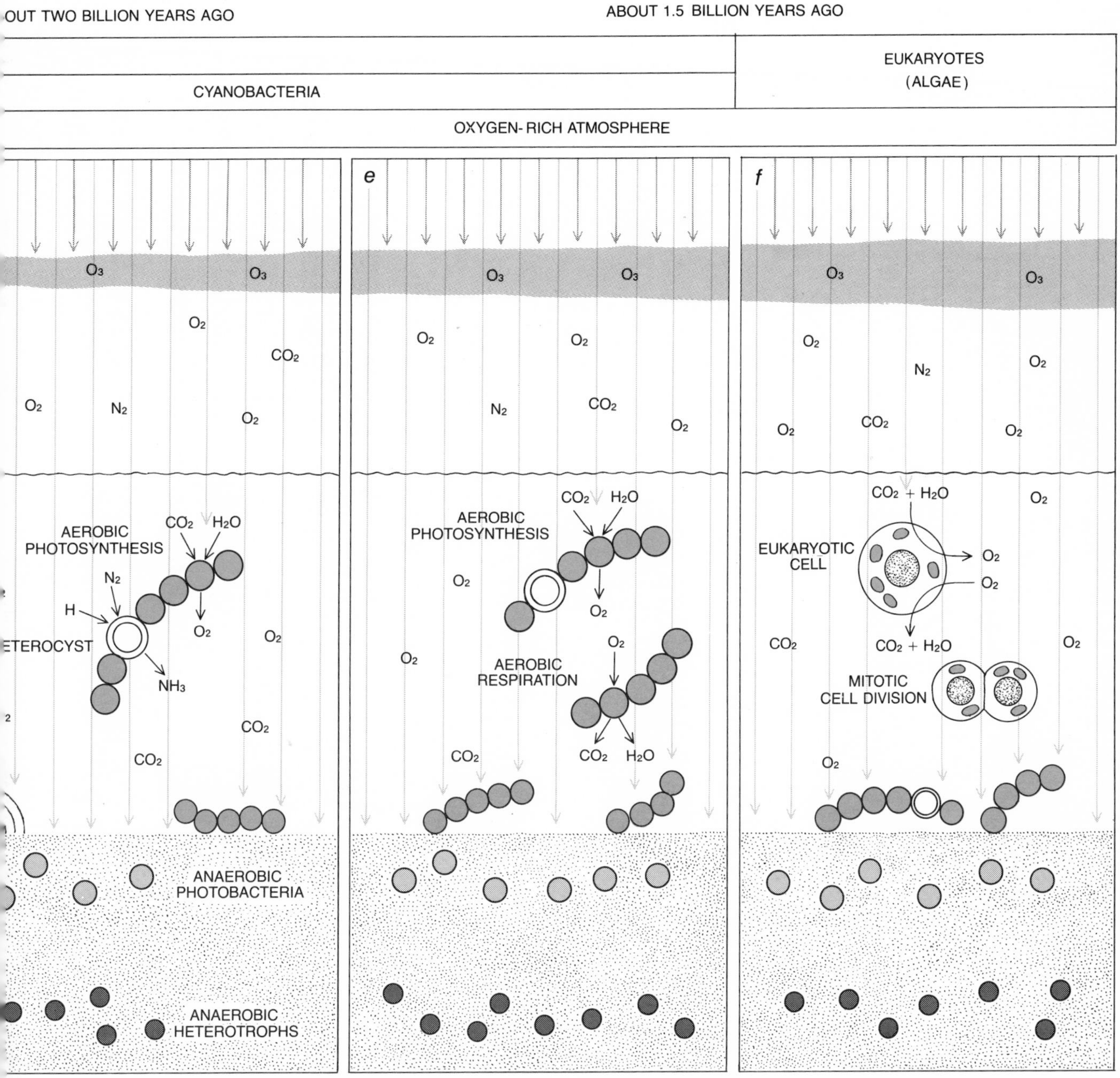

been swept free of iron and similar materials (*d*) did the concentration of free oxygen begin to rise toward modern levels. This biologically induced change in the environment had several effects on biological development. Anaerobic organisms were forced to retreat to anoxic habitats, leaving the best spaces for photosynthesis to the cyanobacteria. In a similar manner nitrogen-fixing organisms had to adopt an anaerobic way of life or develop protective heterocyst cells. Atmospheric oxygen also created a layer of ozone (O_3) that filtered out most ultraviolet radiation. Once the oxygen-rich atmosphere was fully established (*e*) cells evolved that not only could tolerate oxygen but also could employ it in respiration. The result was a great improvement in metabolic efficiency. Finally, about 1,450 million years ago, the first eukaryotic cells emerged (*f*). From the start they were adapted to a fully aerobic environment. The new modes of reproduction possible in eukaryotes, in particular the advanced sexual reproduction that evolved later, led to rapid diversification of the group.

life, some of them recognizable antecedents of modern plants and animals.

In style and in tempo evolution in the Precambrian was distinctly different from that in the later, Phanerozoic era. The Precambrian was an age in which the dominant organisms were microscopic and prokaryotic, and until near the end of the era the rate of evolutionary change was limited by the absence of advanced sexual reproduction. It was an age in which the major benchmarks in the history of life were the result of biochemical and metabolic innovations rather than of morphological changes. Above all, in the Precambrian the influence of life on the environment was at least as important as the influence of the environment on life. Indeed, the metabolism of all the plants and animals that subsequently evolved was made possible by the photosynthetic activities of primitive cyanobacteria some two billion years ago.

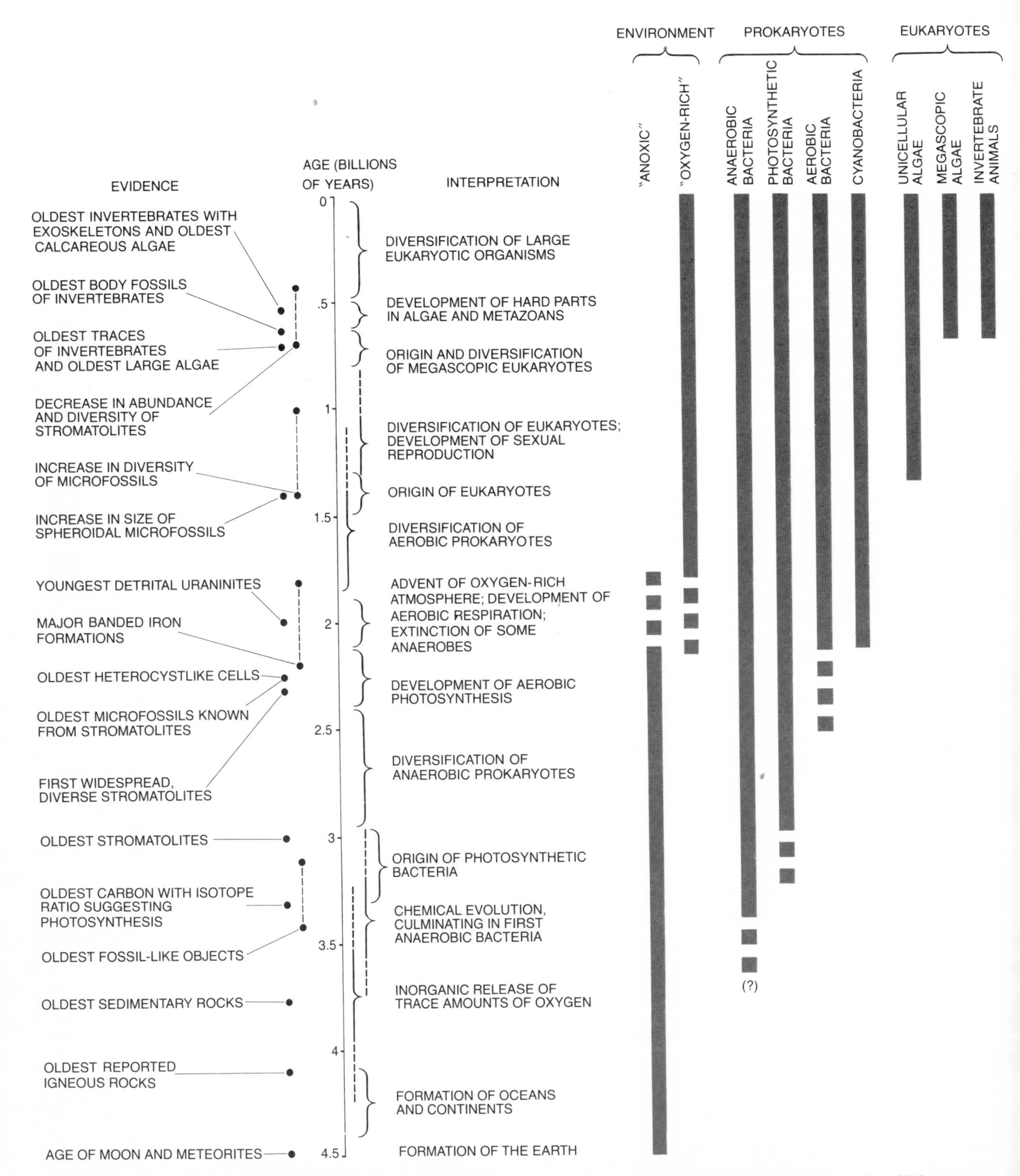

MAJOR EVENTS in Precambrian evolution are presented in chronological sequence based on evidence from the fossil record, from inorganic geology and from comparative studies of the metabolism and biochemistry of modern organisms. Although the conclusions are tentative, it appears that life began more than three billion years ago (when the earth was little more than a billion years old), that the transition to an oxygen-rich atmosphere took place roughly two billion years ago and that eukaryotes appeared by 1.5 billion years ago.

V

The Evolution of Multicellular Plants and Animals

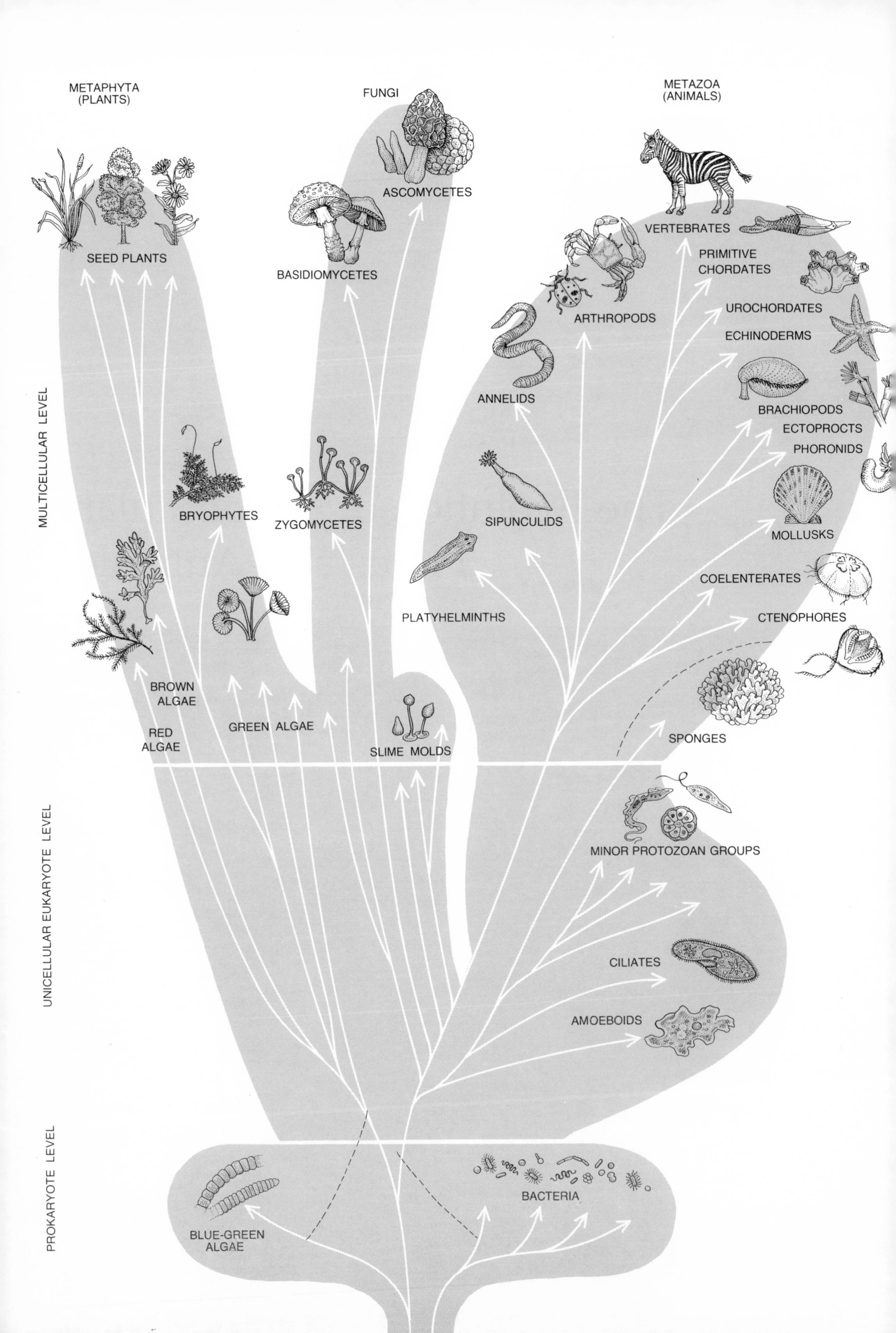

METAPHYTA (PLANTS)
FUNGI
METAZOA (ANIMALS)
ASCOMYCETES
SEED PLANTS
BASIDIOMYCETES
VERTEBRATES
PRIMITIVE CHORDATES
UROCHORDATES
ECHINODERMS
ARTHROPODS
ANNELIDS
BRACHIOPODS
ECTOPROCTS
PHORONIDS
MULTICELLULAR LEVEL
BRYOPHYTES
ZYGOMYCETES
SIPUNCULIDS
MOLLUSKS
COELENTERATES
PLATYHELMINTHS
CTENOPHORES
BROWN ALGAE
RED ALGAE
GREEN ALGAE
SLIME MOLDS
SPONGES
MINOR PROTOZOAN GROUPS
UNICELLULAR EUKARYOTE LEVEL
CILIATES
AMOEBOIDS
PROKARYOTE LEVEL
BACTERIA
BLUE-GREEN ALGAE

The Evolution of Multicellular Plants and Animals

BY JAMES W. VALENTINE

It has been only during the last fifth of the history of life on the earth that multicellular organisms have existed. They appear to have arisen from unicellular organisms on numerous occasions

The animals and plants one sees on the land, in the air and on the water are all multicellular, made up of millions and in some cases billions of individual cells. Even the simplest multicellular organisms include several different kinds of cells, and the more complicated ones have as many as 200 different kinds. All the multicellular plants and animals have evolved from unicellular eukaryotes of the kind described by J. William Schopf in the preceding article. G. Ledyard Stebbins of the University of California at Davis estimates that multicellular organisms have evolved independently from unicellular ancestors at least 17 times. At least two million multicellular species exist today, and many others have come and gone over the ages.

Clearly the multicellular grade of construction is advantageous and successful. The chief advantages of multicellularity stem from the repetition of cellular machinery it entails. From this feature flows the ability to live longer (since individual cells can be replaced), to produce more offspring (since many cells can be devoted to reproduction), to be larger and so to have a greater internal physiological stability, and to construct bodies with a variety of architectures. Moreover, cells can become differentiated (specialized for a particular function, as nerve and muscle cells are), with a resulting increase in functional efficiency. The particular advantages that were the keys to the evolution of multicellularity probably varied from case to case.

The largest categories employed for the classification of organisms are the kingdoms. Multicellular organisms are placed in one or another of three kingdoms on the basis of their broad modes of life and particularly of their modes of obtaining energy. Plants, which are autotrophs (meaning that they require only inorganic compounds as nutrients), utilize the energy of the sun to create living matter through photosynthesis; they make up the kingdom Metaphyta. Fungi (such as mushrooms, which are plantlike but feed by ingesting organic substances) make up the kingdom Fungi. Animals, which are also ingesters, comprise the kingdom Metazoa. Each kingdom includes more than one lineage that evolved independently from the kingdom Protista, consisting of eukaryotic unicellular organisms.

Much of what is known about the evolution of multicellular organisms comes from the fossil record. The Fungi are so poorly represented as fossils that their evolution is obscure, but the other kingdoms have a rich fossil history.

The patterns of adaptation that one can observe today amply demonstrate the effectiveness of evolution in shaping organisms to cope with their environment. Each environment contains animals particularly suited to exploit the conditions in it; each kind of organism has been developed by selection to perform a role in the biosphere. When an environment changes, natural selection acts to change the adaptations; sometimes new environmental roles are evolved. The history of life reflects an interaction between environmental change on the one hand and the evolutionary potential of organisms on the other. It is therefore of interest to briefly examine the major causes of the more biologically important environmental changes.

Notable among them are the processes of plate tectonics. Continents, riding on huge plates of the lithosphere (the outer layer of the solid earth) that move at a rate of a few centimeters per year, break up or collide and can become welded together following a collision. Thus a continent can fragment or grow, the number of continents can increase or decrease and the geographic patterns of a continent can change radically. Ocean basins too alter their sizes, numbers, positions and patterns. The consequences for living organisms can be profound.

Consider only one of the possible events resulting from plate tectonics: the collision of two continents, once widely separated, to form a single larger continent. The accompanying changes in the biological environment are far-reaching. The most obvious change is that the barriers to migration are broken down and the biotas of two continents now compete for existence on one continent. For many land animals the continental interior is now farther from the sea, and the moderating effects of marine air and temperature are diminished. Mountains rising along the suture will diversify the environment further, creating rain "shadows" and perhaps deserts if they happen to interrupt major flows of moisture-laden wind.

Entirely new environmental conditions may thus appear. They create opportunities for new modes of life, as does the general diversification of conditions following the collision. The biotas of the two former continents are subjected to competition and to environmental conditions for which they have not been adapted, and at the same time they are presented with novel opportunities. Evolution can be expected to produce a considerable change in the flora

KINGDOMS OF ORGANISMS are charted according to a concept originated by Robert H. Whittaker of Cornell University. The relatively simple unicellular Monera, which are prokaryotic (lacking a nucleus), gave rise to the more complex unicellular Protista, from which all three multicellular kingdoms have arisen. Multicellular organisms are placed in the kingdoms Metaphyta, Fungi and Metazoa mainly on the basis of the process by which they obtain their energy.

and fauna. The marine organisms of the shallow continental shelves, where 90 percent of marine species live today, will also be affected. For many of them the increased continental area leads to a lowering of environmental stability, requiring new adaptive strategies. In general it is expected that fewer marine species could be supported around one large continent than could be supported around two separate smaller ones.

As oceans widen or narrow, as continents drift into cooler or warmer zones, as winds and ocean currents are channeled in new directions, the pattern and the quality of the environment change. The pace of change may often be slow, because of the low rate of sea-floor spreading, the phenomenon that drives the drifting of the continents and the opening of the oceans. At other times more rapid and dramatic changes can be expected, as for example when continents finally collide after millions of years of approaching each other or when an ocean current is finally deflected from an ancient path.

Just as changes in the environment can affect organisms, so can the activities of organisms affect the environment to create new conditions. A fateful example is the rise of free oxygen in the atmosphere, owing chiefly to photosynthesis. Early organisms could not have existed with free oxygen, whereas most contemporary organisms cannot exist without it. Another point to bear in mind is that organisms form part of one another's environment and interact in numerous ways: as predators, competitors, hosts and habitats. As populations of organisms increase, decrease or change, the environment changes too.

In the course of the diversification of the multicellular kingdoms over the past 700 million years major new types of organisms have appeared and several revolutions have taken place within established groups. In many instances it is possible to identify the kind of environmental opportunity (or, with extinctions, the environmental foreclosure) to which the biota is responding. The history of animals is known best. They arose at least twice: sponges from one protistan ancestor and the rest of the metazoans from another. The major categories into which animals are divided are phyla; over the aeons at least 35 phyla have evolved, of which 26 are living and nine are extinct.

The fossil record that reveals something of the circumstances of the early members of the phyla varies in quality according to the type of fossil. Certain trace fossils are the most easily preserved, particularly burrows and trails left behind in sediments by the activity of organisms. Next come durable skeletal remains, such as seashells and the bones of vertebrates. Finally, the fossils of entirely soft-bodied animals turn up on rare occasions, usually as impressions or films in ancient sea-floor sediments.

The earliest animal fossils are burrows that begin to appear in rocks younger than 700 million years, late in the Precambrian era. Both long horizontal burrows and short vertical ones are found, comparable in size to the burrows of many modern marine organisms. The ability to burrow implies that the animals had evolved hydrostatic skeletons, that is, fluid-filled body spaces that work against muscles, so that the animal could dig in the sea bed. Although some simple animals such as sea anemones manage to employ their water-filled gut as a hydrostatic skeleton and so to burrow weakly, long horizontal burrows suggest a more active animal, probably one with a coelom, or true body cavity. This is quite an advanced grade of organization to find near the base of the fossil record of multicellular forms. Trace fossils are rather rare until about 570 million years ago, when they increase remarkably in kind and number.

The next animal fossils, surprisingly, are soft-bodied remains from between 680 and 580 million years ago, called the Ediacaran fauna from the region in southern Australia where they are best known. The ones that are clearly identifiable with modern phyla are all jellyfishes and their allies, which are at a simple grade of construction. The other fossils are more enigmatic; a few may be allied with living phyla (one resembles annelid worms) and some may not be.

MILLIONS OF YEARS AGO: 0, 50, 100, 150, 200, 250, 300, 350, 400, 450, 500, 550, 600, 650, 700

ERA	PERIOD		EPOCH	EVENTS
CENOZOIC	QUATERNARY		PLEISTOCENE	EVOLUTION OF MAN
	TERTIARY		PLIOCENE MIOCENE OLIGOCENE EOCENE PALEOCENE	MAMMALIAN RADIATION
MESOZOIC	CRETACEOUS			LAST DINOSAURS FIRST PRIMATES FIRST FLOWERING PLANTS
	JURASSIC			DINOSAURS FIRST BIRDS
	TRIASSIC			FIRST MAMMALS THERAPSIDS DOMINANT
PALEOZOIC	PERMIAN			MAJOR MARINE EXTINCTION PELYCOSAURS DOMINANT
	CARBO-NIFEROUS	PENNSYLVANIAN		FIRST REPTILES
		MISSISSIPPIAN		SCALE TREES, SEED FERNS
	DEVONIAN			FIRST AMPHIBIANS JAWED FISHES DIVERSIFY
	SILURIAN			FIRST VASCULAR LAND PLANTS
	ORDOVICIAN			BURST OF DIVERSIFICATION IN METAZOAN FAMILIES
	CAMBRIAN			FIRST FISH FIRST CHORDATES
PRECAMBRIAN	EDIACARAN			FIRST SKELETAL ELEMENTS FIRST SOFT-BODIED METAZOANS FIRST ANIMAL TRACES (COELOMATES)

MAJOR EVENTS in the evolution of multicellular organisms over the past 700 million years are depicted chronologically. The data are based principally on what is revealed by fossils.

SOFT-BODIED ANIMAL of middle Cambrian times left a record in the Burgess Shale of British Columbia. The creature was a polychaete worm with a number of setae, or bristlelike parts, that appear clearly here. The photograph of the fossil, which is enlarged about five diameters, was made in ultraviolet radiation by S. Conway Morris of the University of Cambridge. Lamp was set at a high angle to specimen.

PLANT FOSSILS include a tree fern from Jurassic times (*left*) and a birch leaf from the Miocene epoch (*right*). The ferns were among the earliest land plants, part of the dominant land flora in Devonian times. Birch leaf represents a more advanced group, vascular land plants.

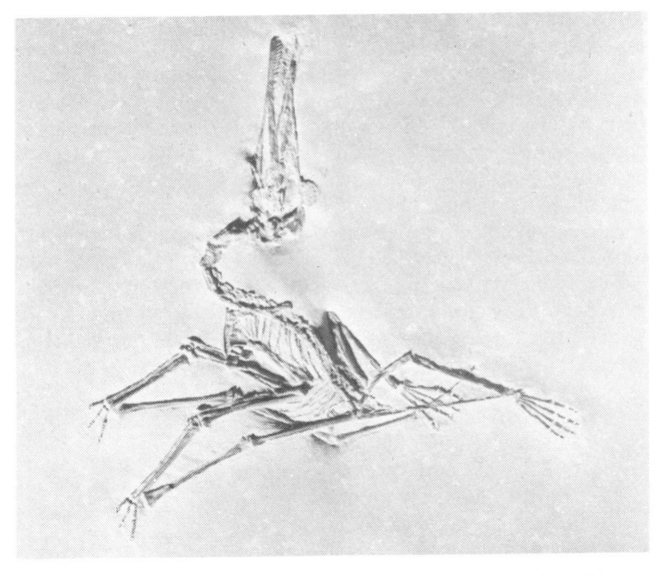

SKELETAL ANIMALS represented in the fossil record include a bed of crinoids (*left*) from the upper Cretaceous period and a pterodactyl (*right*) of late Jurassic times. The crinoids were marine echinoderms, a group represented today by such organisms as starfishes and sea urchins. The pterodactyl was a flying reptile that had a membranous, featherless wing. This specimen was found in Germany.

It is probable that some are coelomates.

Durable skeletal remains finally appear in the fossil record in rocks that are about 580 million years old. The earliest ones are minute scraps, denticles and plates of unknown affinity that were parts of larger animals. Then, starting some 570 million years ago and continuing over the next 50 million years or so, nearly all the coelomate phyla that possess durable skeletons appear in what is in evolutionary terms a quick succession. The exceptions are the phylum Chordata (which nonetheless does appear as a soft-bodied fossil) and the phylum Bryozoa, which finally appears less than 500 million years ago. These durable skeletonized invertebrates seem to have one thing in common: they all originally lived on the sea floor rather than burrowing in it, although one group (the extinct Trilobita) probably grubbed extensively in the sea floor, digging shallow pits or perhaps burrows in search of food.

A highly valuable fossil assemblage from near the middle of the Cambrian period is found in British Columbia in a rock unit named the Burgess Shale. Much of the fauna from this shale is soft-bodied, trapped by rapidly deposited muds and preserved as mineral films by a process that has yet to be determined. Here is an array of more or less normal invertebrate skeletons associated with such soft-bodied phyla as the Annelida (the group that includes the living earthworms), the Priapulidae (possibly pseudocoelomate worms) and the Chordata. The Burgess Shale also contains several animals that represent phyla not previously known. Only one of the phyla could be ancestral to a living phylum; the others are entirely distinct lineages that arose from unknown late Precambrian ancestors and have all become extinct.

From these facts and from the wealth of accumulated evidence on the comparative embryology and morphology of the living representatives of these fossil groups it is possible to build a picture of the rise of the major animal groups. The animals for which the fewest clues exist are the first ones, the founding metazoans, although there is no shortage of speculation on what they may have been like.

Since the evolution of novel organisms involves adaptation to new or previously unexploited conditions, one can imagine the adaptive pressures the earliest multicellular forms would have faced and can thereby derive a variety of plausible animals radiating from a trunk lineage into the available modes of life. Bottom-dwelling animals can eat deposited or suspended food. Suspension feeders must maximize the volume of water on which they can draw. Body shapes such as domes that jut up into the water and create turbulence in the ambient currents would increase the volume of water washing the animal and would be advantageous. In quiet waters taller cylinders or extensions of the body, perhaps in the form of filaments or tentacles, would increase the feeding potential. For deposit feeders flattened shapes would maximize the area of contact with the bottom.

A differentiation of function among cells would be an early trend among such animals: In a deposit feeder, for example, the cells on the bottom would be ingesting food and so could most easily specialize for digestion, perhaps becoming internalized to increase their number and stabilize the digestive processes. Covering cells would be supportive and protective. Marginal cells could specialize for locomotion. Well-nourished cells surrounding the digestive area might specialize in reproduction.

Patterns of differentiation would vary in organisms of other shapes. In quiet-water suspension feeders the digestive cells might lie uppermost. Floating animals would have other shapes, perhaps radial or globose. Any such simple animal type could have been ancestral to the remaining phyla. Such animals are essentially at the same grade as sponges, but not one of them survives today. (Sponges are so distinctive in their developmental pattern that they probably arose from unicellular organisms independently of all other animal phyla.)

It is impossible to estimate how long before 700 million years ago the first animals evolved; it could easily have been as little as 50 million years or as much as 500 million. At any rate the evolutionary trends of differentiation in organs and tissues led to the rather complex invertebrates, most or all of which arose as burrowers in the sea floor. From the body plan of living phyla that first appeared during the Cambrian period one can infer the types of body plan that must have been present among late Precambrian animals. For example, two of the most important types of coelomates in Cambrian times were metamerous and oligomerous forms.

In the metamerous type the coelom is divided by transverse septa into a large number of compartments, as it is in earthworms. The muscular activity associated with burrowing affects only the segments in the immediate vicinity of the contractions, so that the efficiency of burrowing is enhanced. The oligomerous body plan has only three (sometimes only two) coelomic compartments separated by transverse septa. Each coelomic region functions differently. In a particularly primitive body type, represented by living phoronid worms, the regions correspond to a long trunk in which the coelom is used in burrowing and to a tentacular crown that is employed in feeding.

It seems that between the development (perhaps more than once) of the coelom about 700 million years ago and the appearance of animals with durable skeletons about 570 million years ago coelomate worms diversified considerably within their adaptive zones. Most of these groups lived in burrows within soft sediments of the ancient sea floor. Then, beginning some 570 million years ago, a rich fauna with durable skeletons appeared. The animals lived on the sea floor, where conditions are considerably different from those within it. Most of the lineages were modified in adapting to this new zone. Animals that descended from oligomerous suspension feeders tended to evolve skeletons that served to protect and support their feeding apparatus. Some of the more mobile burrowers, such as metamerous worms, developed a jointed external skeleton with jointed appendages, which operated largely as a system of levers. Their living descendants include insects, crabs and shrimps (phylum Arthropoda).

In these and other instances the evolution of a durable skeleton was associated with a large number of coadapted changes in the anatomy of the soft parts. For example, the arthropod coelom was reduced, since the exoskeleton took over locomotion and limbs replaced the wavelike action of peristalsis in achieving movement. Septa between segments were no longer required, and they disappeared. The original metamerous architecture is still evidenced by a regular series of paired internal organs, inherited from a segmented ancestry.

All the evidence points to a major phylum-level diversification near the beginning of the Paleozoic era. Phyla that still exist appeared, along with the extinct soft-bodied phyla represented in the Burgess Shale. Some early skeletal types that are now extinct may also have been distinctive phyla. One phylum resembled sponges, and another may have been allied to mollusks. At least a third again as many phyla as are alive are known from Paleozoic fossils, and there must have been even more.

The environmental context of this radiation is not understood. It has been

LIVING ANIMAL PHYLA are grouped on the opposite page according to their body architecture, with a representative of each phylum portrayed. The groups are (*a*) simple multicellular forms in which a single tissuelike layer surrounds a central cavity; (*b*) forms with two tissue layers, one of which surrounds a well-defined gut; (*c*) wormlike forms with three tissue layers, the middle one forming the core of the body surrounding the gut; (*d*) small, usually parasitic forms with three tissue layers and a primitive body cavity, and (*e–h*) four groups with "true" body cavities that are insulated from the external environment; the form of the cavity differs somewhat from group to group. There are 26 living phyla and at least nine extinct ones.

PORIFERA (SPONGES)

a

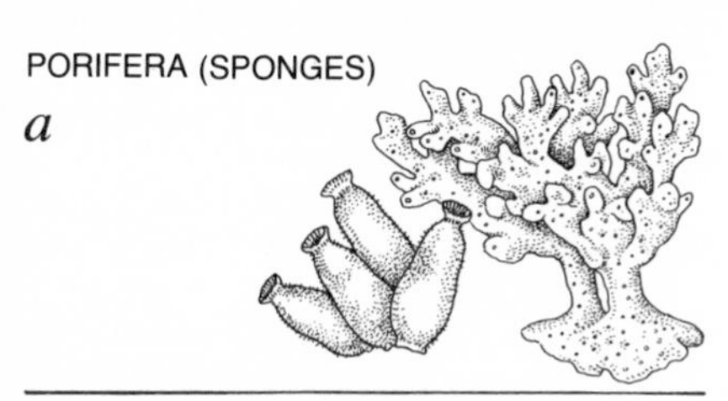

CNIDARIA (SEA ANEMONES, JELLYFISHES, CORALS)

b

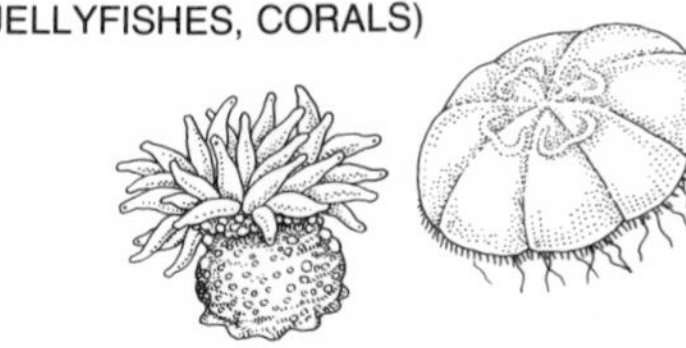

CTENOPHORA (COMB JELLIES)

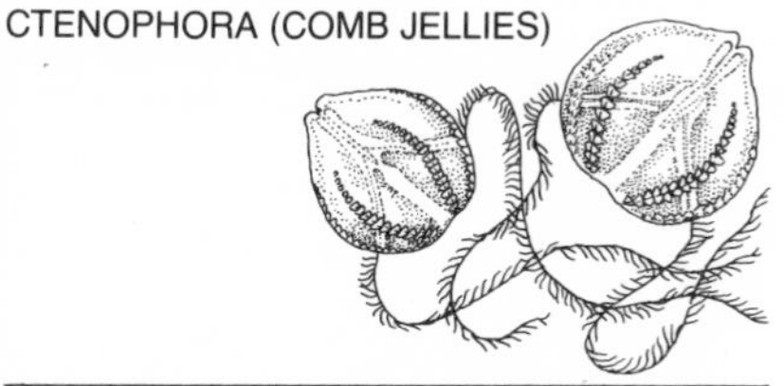

PLATYHELMINTHES (FLATWORMS)

c

NEMERTINA (PROBOSCIS OR RIBBON WORMS)

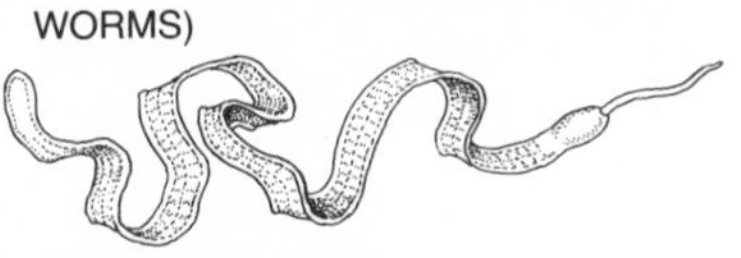

ACANTHOCEPHALA (SPINY-HEADED WORMS)

d

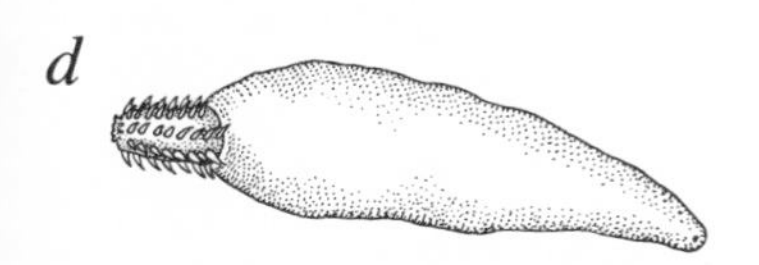

ROTIFERA (WHEEL ANIMALS)

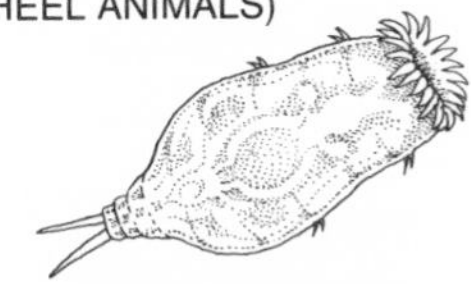

GASTROTRICHA (SCALED WORMS)

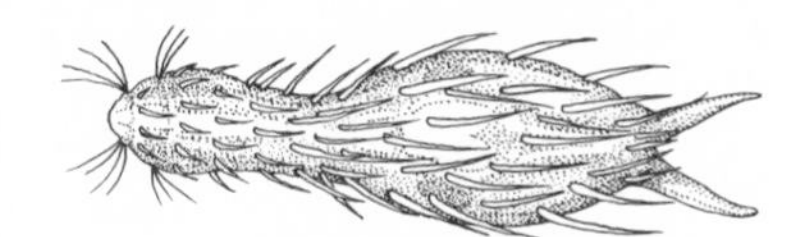

KINORHYNCHA (SPINY-SKINNED WORMS)

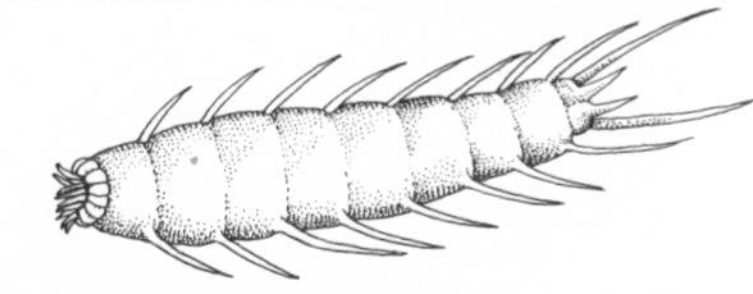

NEMATODA (ROUNDWORMS)

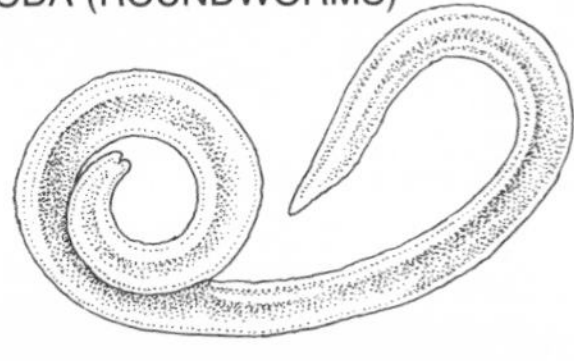

PRIAPULIDA (PRIAPUS WORMS)

d

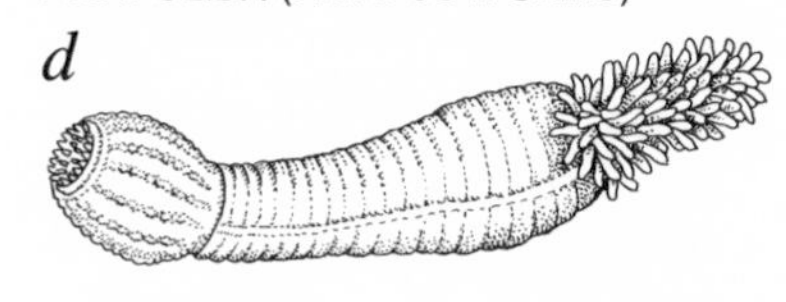

ENTOPROCTA (TENTACLED, STALKED ANIMALS)

ANNELIDA (EARTHWORMS, FAN WORMS)

e

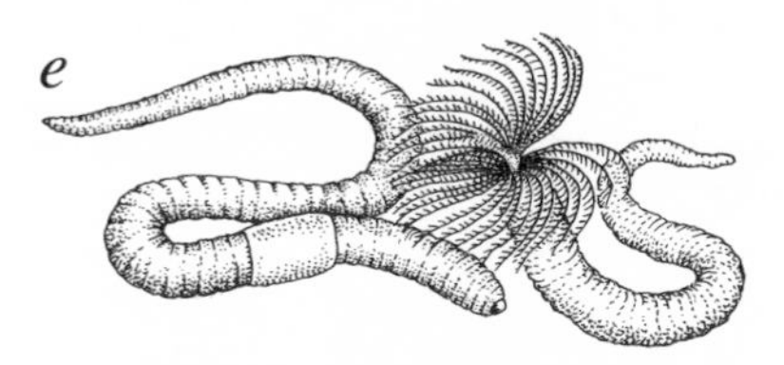

ARTHROPODA (INSECTS, CRABS, SHRIMPS, BARNACLES)

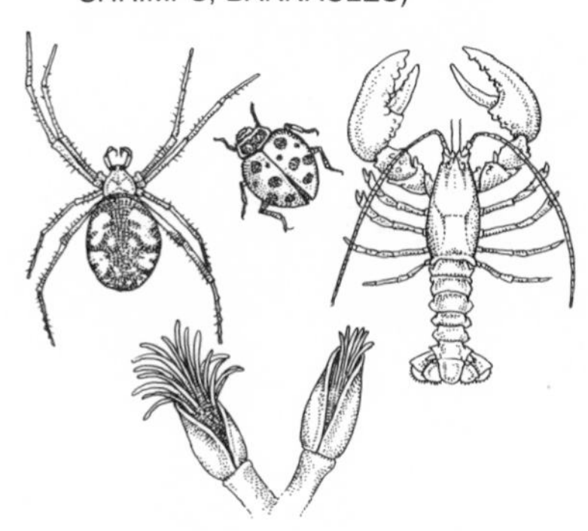

SIPUNCULIDA (PEANUT WORMS)

f

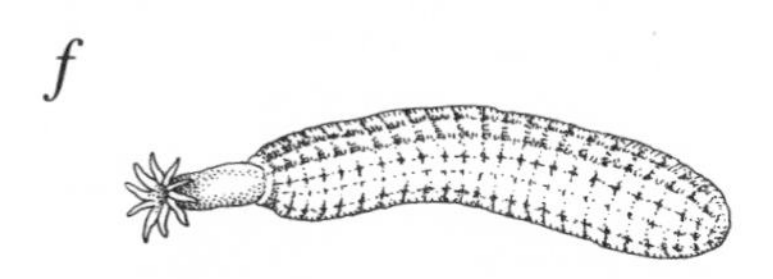

ECHIUROIDEA (SAUSAGE-SHAPED MARINE WORMS)

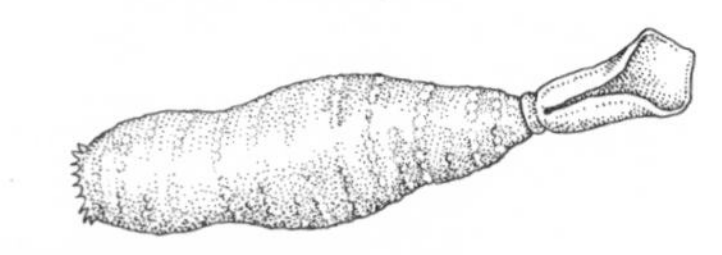

MOLLUSCA (CLAMS, SNAILS, OCTOPUS, SQUID)

g

PHORONIDA (HORSESHOE WORMS)

h

BRACHIOPODA (LAMPSHELLS)

h

ECTOPROCTA (BRYOZOA OR MOSS ANIMALS)

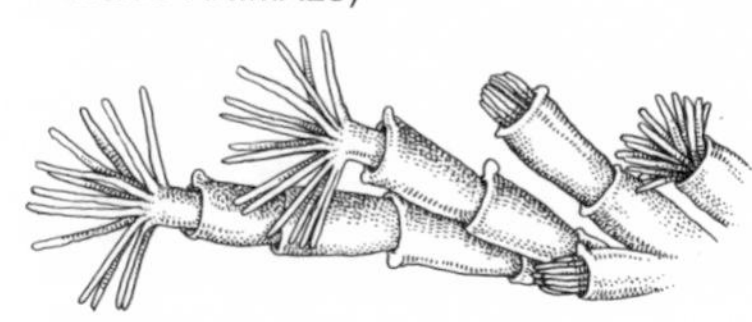

CHAETOGNATHA (ARROWWORMS)

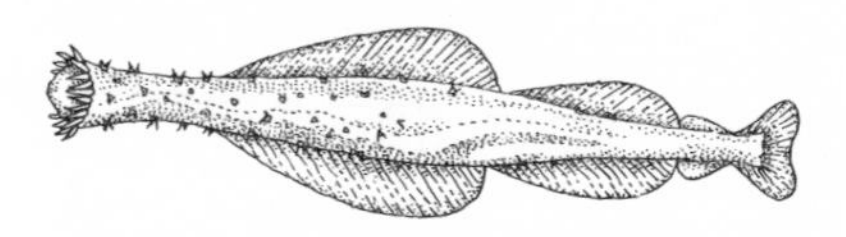

POGONOPHORA (DEEP-SEA WORMS)

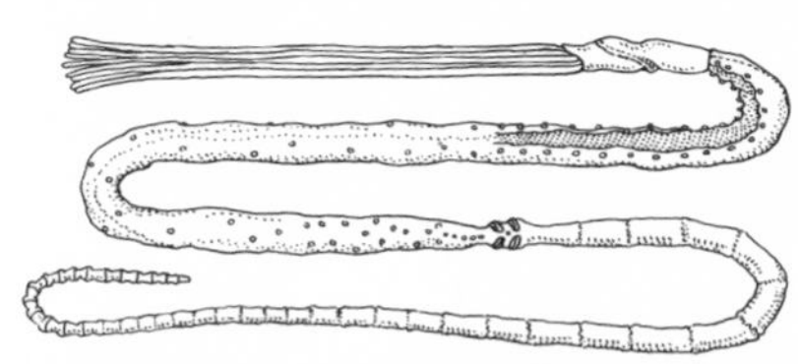

ECHINODERMATA (STARFISHES, SEA URCHINS, SAND DOLLARS)

HEMICHORDATA (ACORN WORMS)

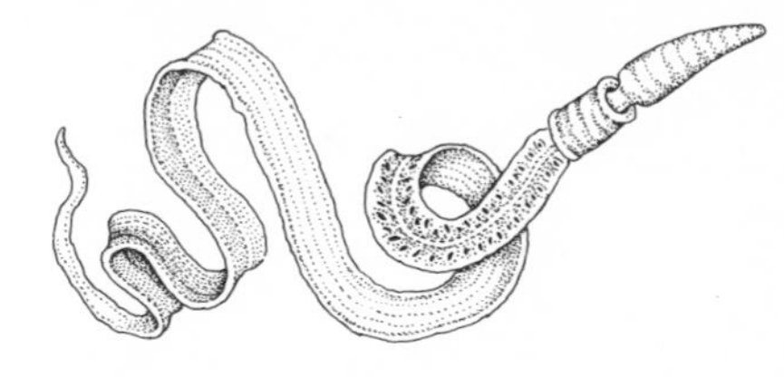

UROCHORDATA (SEA SQUIRTS)

CHORDATA (AMPHIOXUS, FISHES, AMPHIBIANS, REPTILES, BIRDS, MAMMALS)

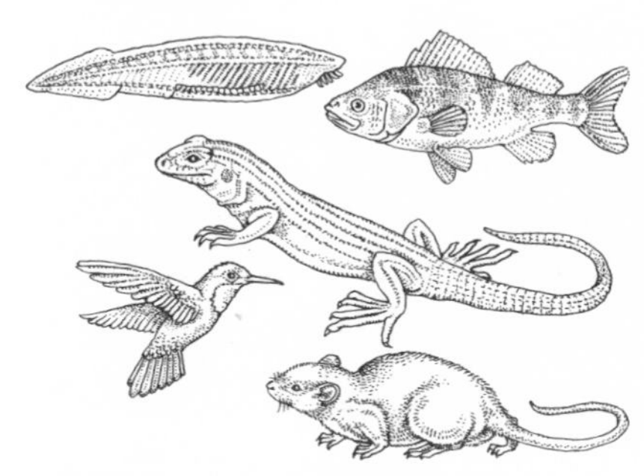

suggested that oxygen may finally have reached a level sufficiently high to support active animals, stimulating an evolutionary burst. Another idea is that the environment became stabler.

Among the many lineages evolving novel modes of life at the time was one that developed a swimming specialization and diverged significantly from other animal groups. This was a suspension-feeding oligomerous animal that evolved a stiffening but flexible dorsal cord and characteristic chevron-shaped blocks of muscle that could flex the cord from side to side for swimming. Eventually the animals developed durable parts: an outer armor of plates for protection and an axial skeleton of articulated vertebrae with elongated lateral flanges to support the body walls. These were the earliest fishes. They were jawless and had unpaired fins. They fed by taking in water through an anterior mouth, ingesting suspended material as

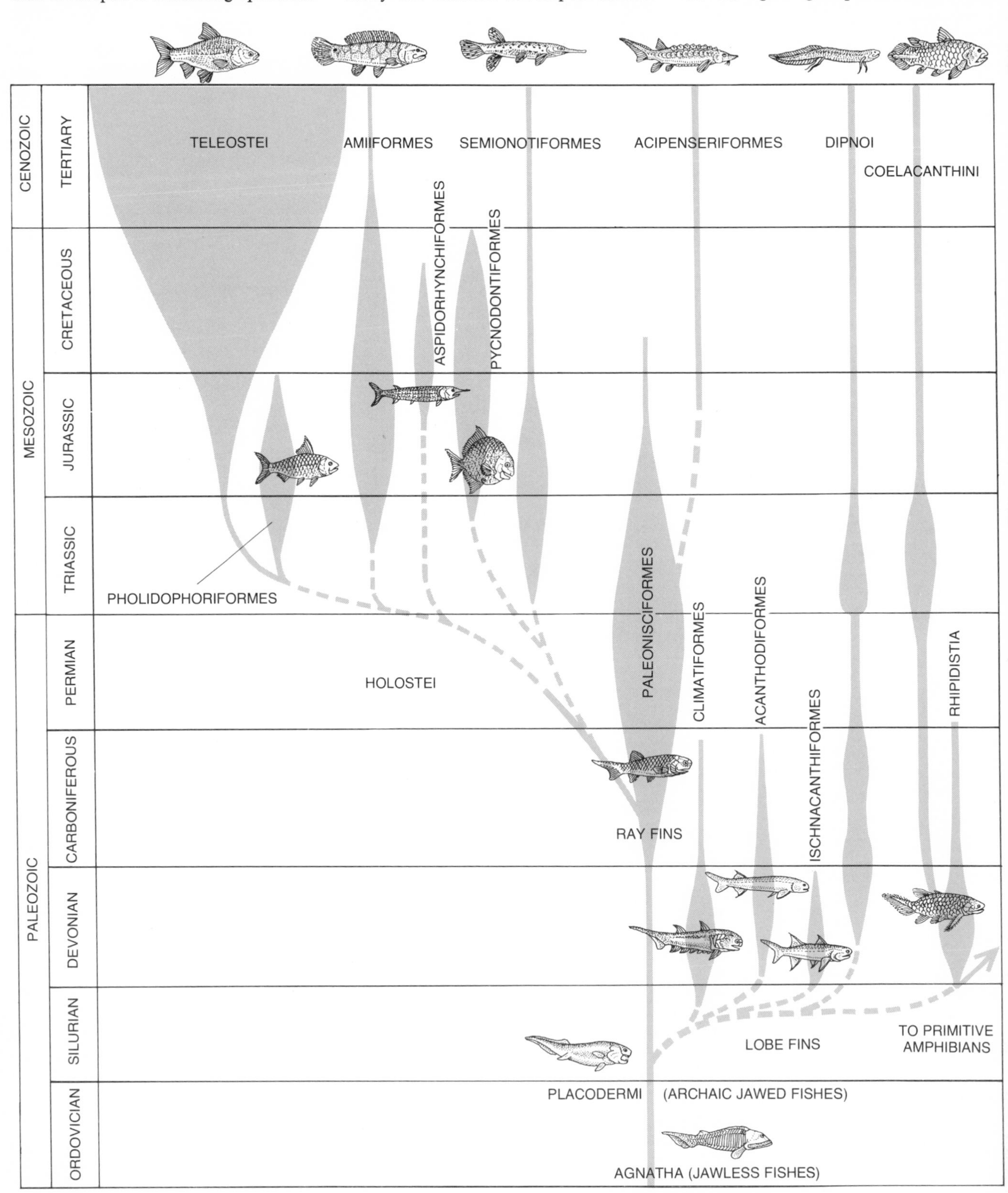

EVOLUTION OF FISHES began in the Cambrian period with the appearance of jawless species that had evolved from a simpler multicellular marine organism. In the Devonian period jaws evolved from a pair of gill slits, and fins became paired. One major line of jawed fishes, the ray fins, was ancestral to most of the fish species now living. The other line, the lobe fins, did not fare as well but led eventually to the earliest amphibians, the ancestors of all four-footed vertebrates. In large part the ability of the early amphibians to venture onto the land was due to the evolution there of multicellular plants that sustained them, giving rise later to animals that lived entirely on land.

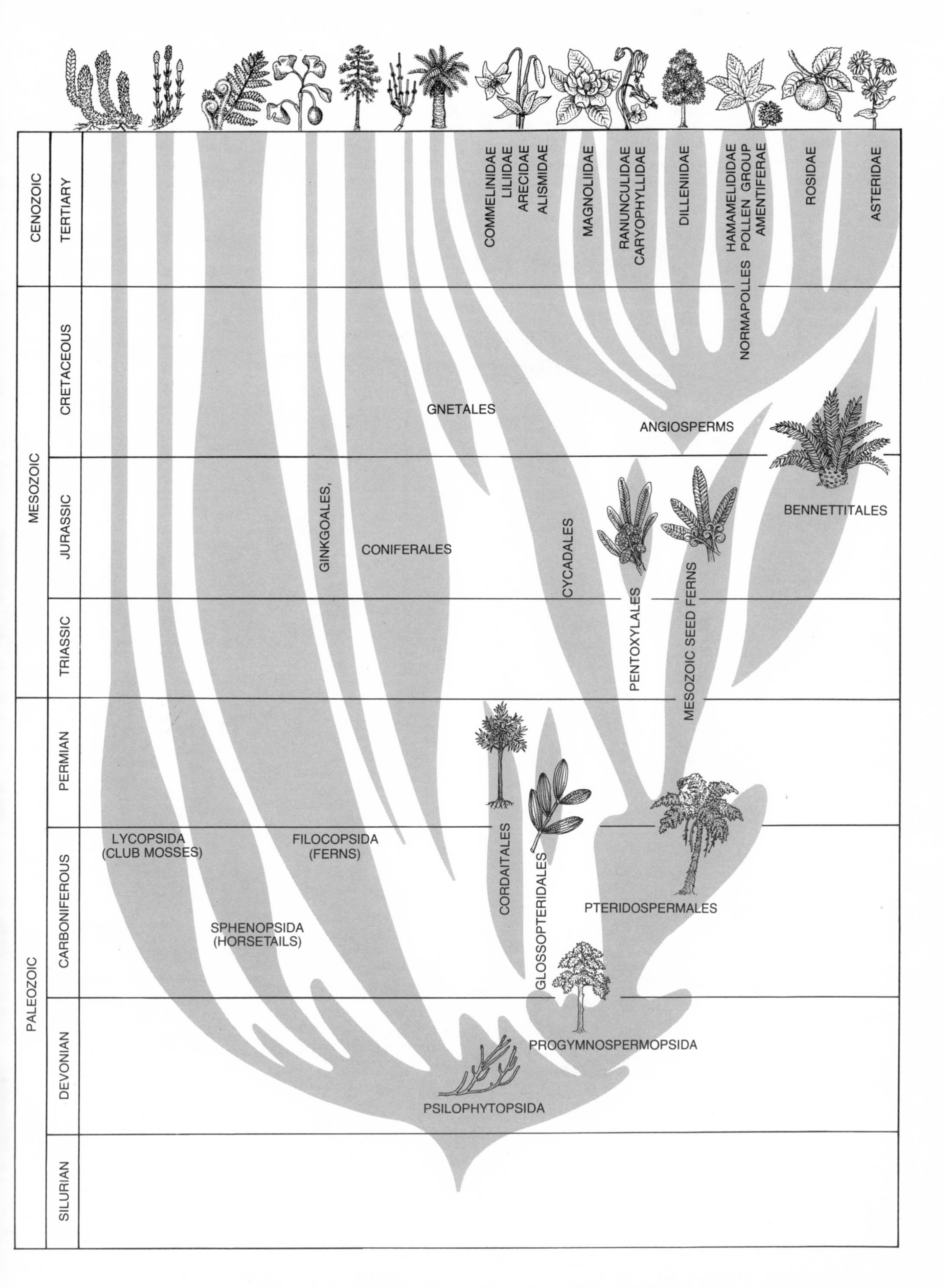

VASCULAR LAND PLANTS probably evolved according to the pattern indicated in this diagram. In the Devonian period the primitive assemblage of horsetails, club mosses and ferns dominated the flora. Those forms reproduce by spores and prefer humid conditions. Seed- and pollen-bearing plants developed by Devonian times, and by the Permian the conifers had begun an expansion that made them dominant in the Mesozoic era. Late in the Cretaceous period the angiosperms (flowering plants) spread explosively and became dominant.

food and expelling the filtered water through their gills.

The early jawless fishes, the agnathans, are first known from late Cambrian fossils. They continued as a successful group into the middle of the Paleozoic era, but they diversified only modestly. At some time in the Devonian period two developments revolutionized the fishes: jaws evolved from the anterior pair of gill slits and fins became paired. With a greatly expanded choice of food items and an improved swimming balance, the jawed fishes diversified spectacularly. (The agnathans declined, although they may still be represented by such forms as the lampreys.)

Two major types of jawed fishes were the ray fins and the lobe fins. The great majority of living fishes have descended from the ray fins. The lobe fins were far less successful as fishes (they survive only as lungfishes and a

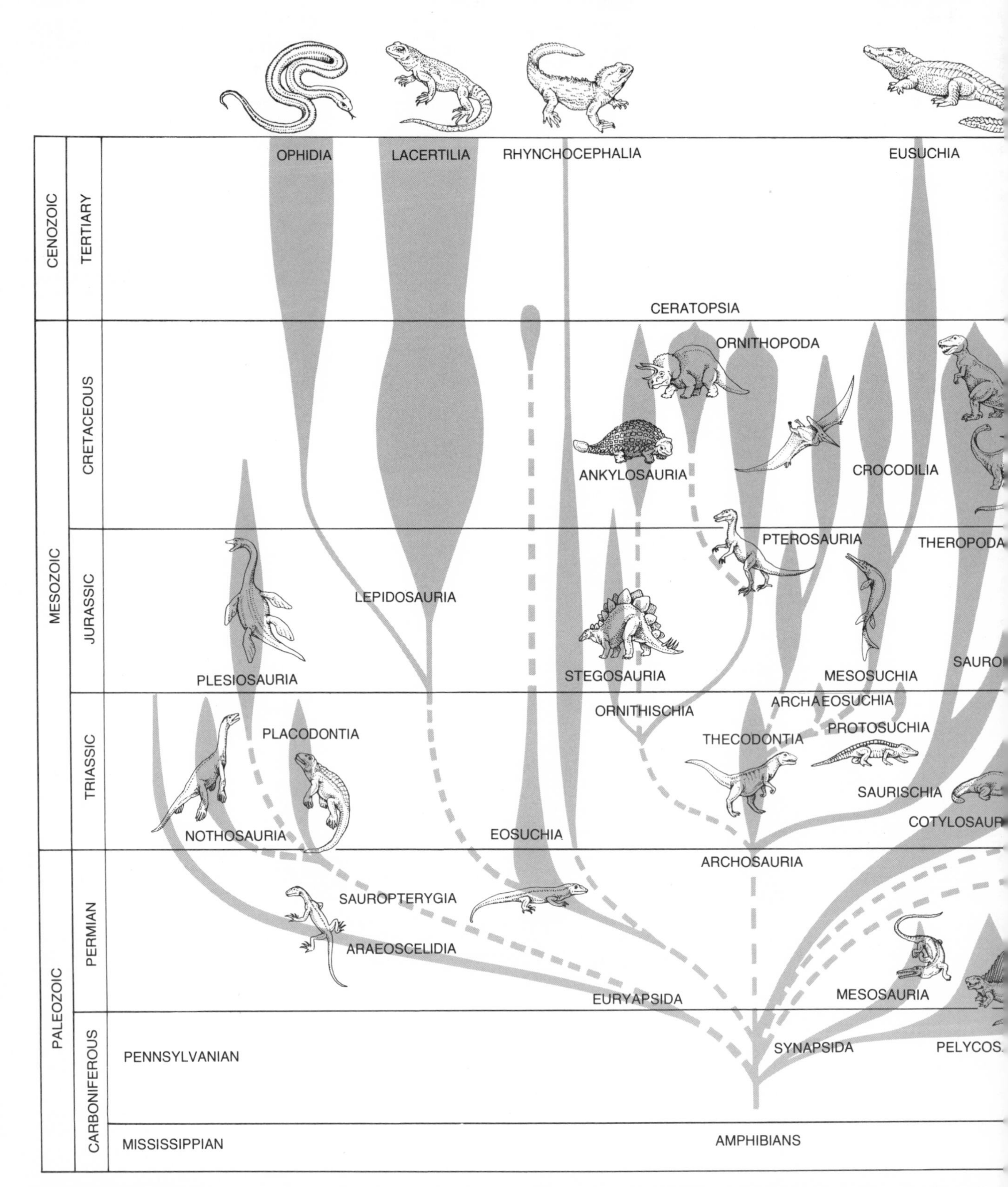

REPTILES EVOLVED from a primitive amphibian stock, which became free from ties to the water by evolving an egg that could develop in a terrestrial setting. Diversifying rapidly, the reptiles eventually came to dominate life on the land. The pelycosaurs and the therapsids were two particularly successful reptile groups, the therapsids gradually becoming the more diversified. In the Triassic period the dinosaurs evolved in the reptile line and began their 150 million years of dominance of the terrestrial environment. In this chart and in the

few relict marine forms), but they had bony supports within their fins from which limbs evolved. The earliest amphibians arose from a primitive group of lobe fins (the rhipidistians), and so all four-footed vertebrates (tetrapods) and their descendants also evolved from this vanished fish stock.

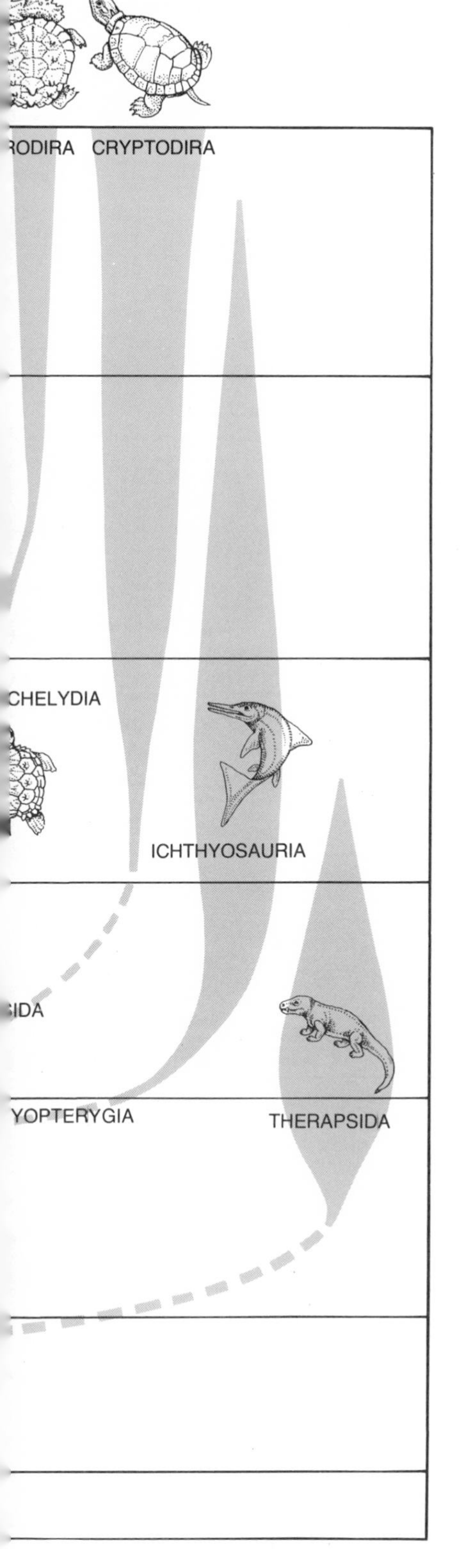

other accompanying charts tracing the evolution of animal and plant groups the width of a line or a band indicates relative size of that group at the corresponding geological period.

The energy to support the early marine animals must have been supplied by photosynthesizing unicellular organisms at first (and it still is supplied chiefly by such forms), but at some time in the late Precambrian era multicellular marine algae evolved. Little is known of the diversity or abundance of these plants; they must have contributed detritus to invertebrate communities and may have been grazed directly. It is possible that soil bacteria, fungi and lower plant forms had colonized the land by the Cambrian period, and perhaps the fringes of swamps and embayments supported hardy, semiaquatic plant types. The first nonaquatic plant lines whose descendants form the major elements in terrestrial flora arose, however, in the Silurian period. The early plants spread from marshes and swamps into drier upland habitats. As the green belt expanded, animals followed it ashore: arthropods and probably worms, feeding on plant debris and eventually on the plants themselves. Thus prey items existed on land to support the populations of larger tetrapods that appeared during the Devonian period.

Many of the early amphibian lineages developed rather large body sizes and radiated into the available habitats, becoming herbivores and predators on many food items in aquatic, semiaquatic and terrestrial settings. Although they must have been quite hardy, perhaps rivaling reptiles in this respect, they were still bound to the water for reproduction, as frogs and salamanders are today. Modern amphibians are quite unlike the large forms that ruled the terrestrial domain for some 75 million years. The modern antecedents probably arose during the late Paleozoic era: small-bodied forms adapted to marginal habitats and utilizing resources not utilized by their larger relatives. Perhaps in this way they escaped competition with later vertebrates.

Reptiles arose from a primitive amphibian lineage, freed from ties to the water by the evolution of an egg that could develop in a terrestrial environment. The reptiles diversified rapidly and spread into all the environments occupied by their large amphibian cousins, becoming successful predators and competitors. The large amphibians declined to extinction late in the Triassic period. Even by late Permian times reptiles were well on the way to dominance. Two groups were particularly successful: the pelycosaurs (known by their large dorsal fins) and the therapsids, which may have been more active and aggressive in view of the fact that they eclipsed the pelycosaurs in diversity early in the Triassic.

The therapsids were replaced in turn as dominant reptiles by the dinosaurs, which evolved in Triassic times and did not become extinct until 150 million years later. During this interval they underwent several severe waves of extinction, which carried off the larger species disproportionately, but each time the dinosaurs reradiated from the surviving smaller stocks to maintain their ascendancy. Finally they disappeared at the close of the Cretaceous period about 65 million years ago.

During most of their tenure the dinosaurs shared the terrestrial world with a group of small, active, hairy animals that evolved from a predatory therapsid lineage: the mammals. The evolution of the mammals is particularly well recorded by fossils, which display the gradual appearance of mammalian skeletal features from reptilian ones. Unfortunately many mammalian characteristics cannot ordinarily be determined from fossils. They include warm-bloodedness, hair, a respiratory diaphragm, increased agility and facial muscles that allow suckling. The mammals must have shared at least some of these features with their therapsid ancestors. In mammals they form a unique adaptive assemblage.

Once the dinosaurs became extinct, mammals radiated into the vacant habitats to dominate in their turn the terrestrial environment. Since mammals are alert and active compared with living reptiles, it is puzzling that they did not radiate sooner to challenge the dinosaurs. A possible reason is that the dinosaurs themselves were active, alert and warm-blooded. The posture of dinosaurs suggests agility; their bones, compared with the bones of warm-blooded and cold-blooded tetrapods alive today, contain channels that suggest warm-bloodedness, and the fossil record of predator-prey ratios suggests that dinosaurs required a large amount of food, like warm-blooded mammals and unlike cold-blooded reptiles.

If dinosaurs were warm-blooded, it would help to explain their long dominance. The dinosaurs were larger than their mammalian contemporaries; even dinosaur hatchlings were larger than most mammals. Clearly the mammals played the smaller-animal roles and the dinosaurs the larger-animal ones. Even when repeated extinctions reduced the diversity of the dinosaurs, the survivors were still larger than the mammals and were able to hold their own and to furnish the basic lineages from which evolution re-created large animals.

The cause of the final extinction of the dinosaurs remains a mystery, but it appears to be related to the carrying capacity of the Cretaceous environment. Only small animals weighing no more than about 20 pounds survived the wave of terrestrial extinctions that closed the Cretaceous period; the mammals were smaller still.

During their long coexistence with the dinosaurs the mammals developed improvements that have stood them in

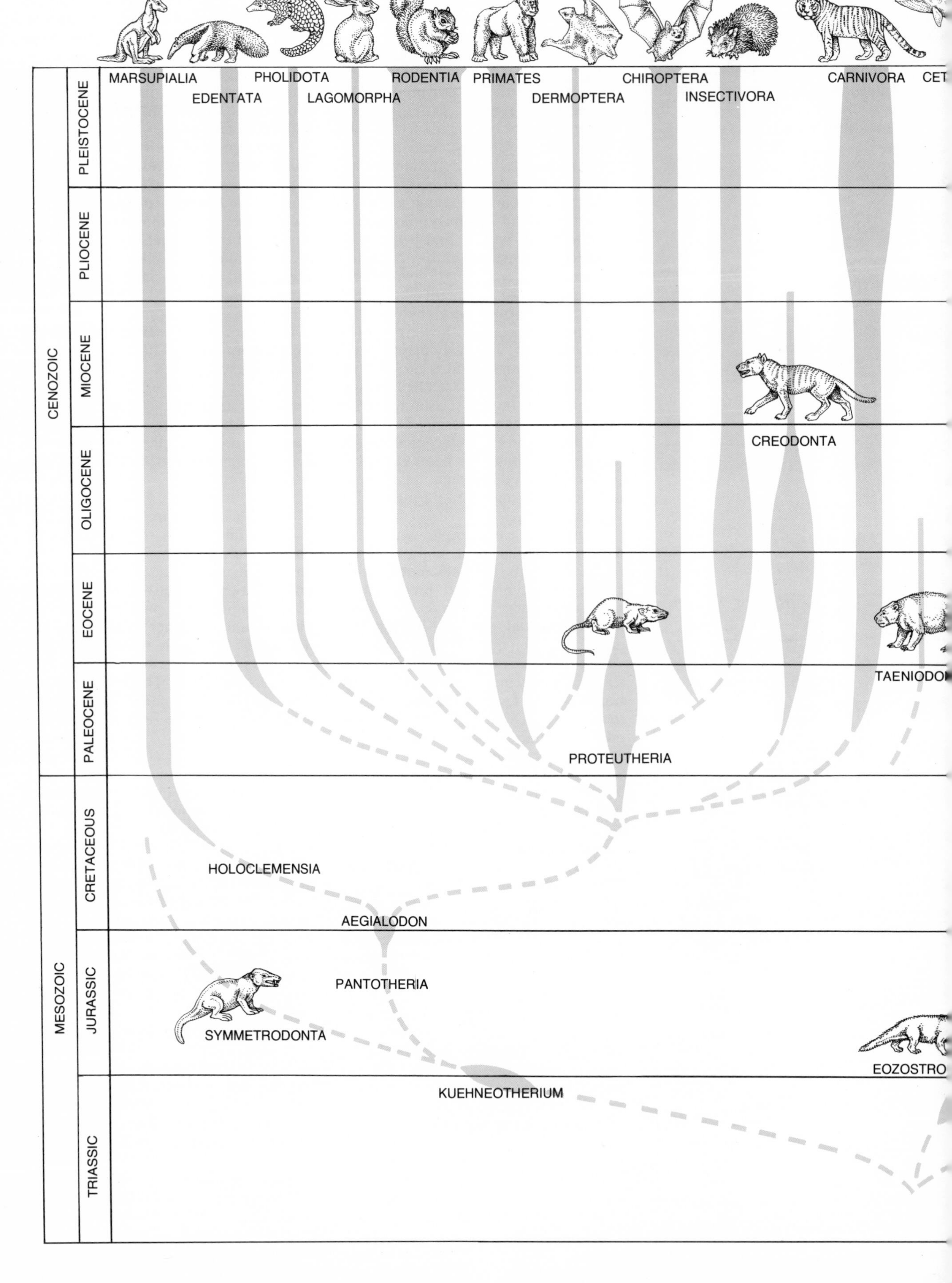

MAMMALIAN EVOLUTION originated in a group of predatory therapsids. The first mammals were quite small; the ones that survived the heavy extinction of land animals that came at the end of the Cretaceous period all weighed less than 20 pounds. Several lines of mammals survived the dinosaurs, and during the Cenozoic era the mammal group diversified widely. The first primates appeared while

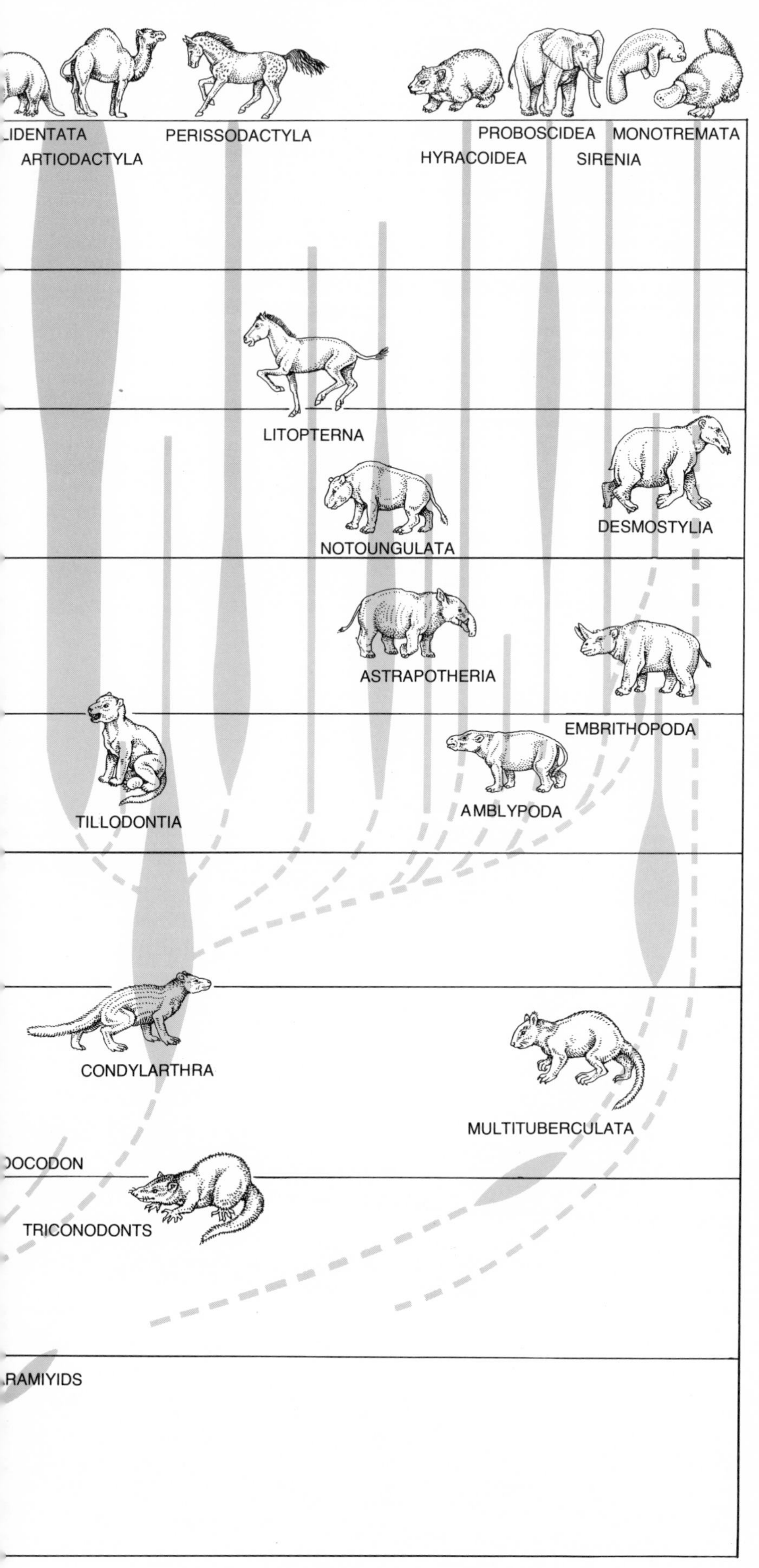

the dinosaurs were still dominant; those primates were small, tree-dwelling creatures with habits probably resembling those of squirrels. The mammal line that led eventually to man took up life on the ground, probably only as foragers at first but later as highly successful hunters.

good stead. Mammalian forms with a placenta arose during the Cretaceous period and diversified moderately. Primates, the order including man, appeared while the dinosaurs still dominated. Hence several mammal lineages survived the dinosaurs. During the Cenozoic era they diversified impressively. A number of lineages that originated then have become extinct, so that fewer orders exist today than were living at times during the Cenozoic era.

The evolutionary activity of the mammals reached its peak within the past two million years, perhaps because of the great climatic diversity associated with the late Cenozoic glaciation. The latest mammalian episode has been a wave of extinction that was particularly severe for large mammals, including manlike species.

The primates of early Cenozoic times were small and probably squirrel-like in their habits. Many characteristic primate features, such as overlapping binocular visual fields, a short face, grasping forepaws and increased brain size and alertness, are probably adaptations to an arboreal existence. The lineage that led eventually to man descended from the trees to the forest floor to forage and eventually to hunt, perhaps coming to live at the margins of forests with access to moderately open country. A continuing adaptation to a terrestrial habitat led to an erect posture. Hunter-gatherer bands developed; perhaps the distinctive tooth arrangement of human beings, with its reduced canine teeth, was associated with dietary shifts as this kind of social evolution proceeded. It has been suggested that the final rise of the human species was associated with a further shift to big-game hunting, increasing the value of cunning, intelligence and cooperation.

The pattern of the history of land plants is similar to that of land vertebrates, with waves of extinction and replacement and the episodic rise of new forms to dominance. In the Devonian period, when the early forests appeared, the primitive assemblage of horsetails, club mosses and ferns spread and came to dominate the land flora. Plants of this form reproduce by spores and prefer humid conditions.

Plants bearing seeds and pollen developed as early as Devonian times. They diversified during the Carboniferous period; by the Permian one lineage, the conifers, began an expansion that led them to dominate the Mesozoic floras. The shift to conifers was associated with the appearance of drier climates.

Still another shift came late in the Cretaceous period, when flowering plants (angiosperms) spread explosively to conquer the terrestrial realm. (About a quarter of a million species of angiosperms are living today.) The earliest flowering plants seem to have been weedy, opportunistic species adapted for rapid reproduction. The reproductive specializations, including the development of flowers and the appearance of insect pollinating systems, were transformed into a general advantage over the more slowly growing conifers.

The details of the diversity and abundance of plant species through the Paleozoic and Mesozoic eras are largely unknown. The major transitions in dominant floral elements resemble what was happening to land animals, but as far as one can tell they do not correspond to the events that were affecting the animals. For example, the angiosperms were well established long before the dinosaurs were extinguished. Moreover, the several waves of extinction of tetrapods during the Mesozoic era are not reflected in the history of the land plants as it is now known.

The evolutionary history of animals and plants, from the rise of the kingdoms and their principal subdivisions to the origin of recent species, appears as a series of biological responses to environmental opportunities. Early diversifications near the beginning of the Cambrian period produced a series of animal body plans, each adapted to a particular mode of life. Many of these plans proved to be preadapted to further life modes and were extensively diversified; today they are what are called phyla. Others became extinct sooner or later. Extinctions are in some ways a measure of the success of evolution in adapting organisms to particular environmental conditions. When those conditions vanish, the organisms vanish. Opportunities are thus created for selection to develop novel kinds of organisms from among the survivors. The new forms are sometimes spectacularly successful, particularly when their adaptations provide entry into a relatively empty niche. The groups that disappear are not replaced by totally new groups but by branches from the remaining lineages. Therefore as time goes by the number of distinctive kinds of organisms declines, whereas the remaining groups become on the average more diverse.

The groups that disappeared were not necessarily less well adapted than or structurally inferior to the survivors, except in relation to the sequence of environmental changes that happened to occur. This sequence, ultimately controlled by physical changes within a dynamic earth, is unrelated to the life forms clinging to adaptation as the means of survival on the continents and in the oceans. If the environmental changes had been different, different survivors would have emerged. One cannot say what life forms would now inhabit the earth if the physical history of the planet had been different. One can only be certain they would differ from the ones that are here today.

VI

The Evolution of Ecological Systems

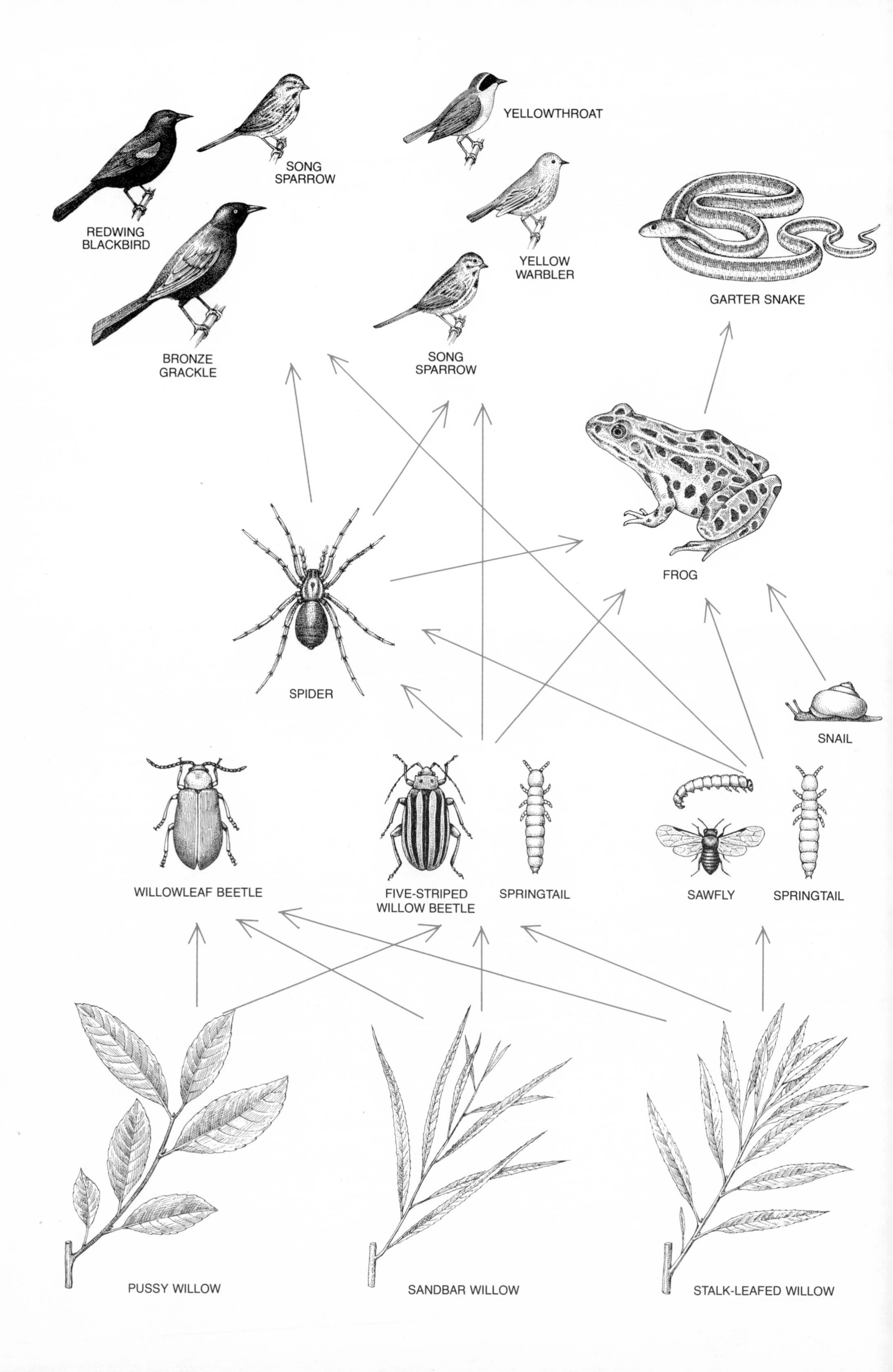
SONG SPARROW
YELLOWTHROAT
REDWING BLACKBIRD
YELLOW WARBLER
GARTER SNAKE
BRONZE GRACKLE
SONG SPARROW
FROG
SPIDER
SNAIL
WILLOWLEAF BEETLE
FIVE-STRIPED WILLOW BEETLE
SPRINGTAIL
SAWFLY
SPRINGTAIL
PUSSY WILLOW
SANDBAR WILLOW
STALK-LEAFED WILLOW

The Evolution of Ecological Systems

BY ROBERT M. MAY

The coevolution of species within ecosystems gives rise to interesting patterns in the total number of species found in a given area, in their relative abundances and in the overall structure of the local food web

Strictly speaking, ecological systems as such do not evolve. As has been stressed repeatedly in this issue, natural selection acts almost invariably on individuals or on groups of related individuals. Populations, much less communities of interacting populations, cannot be regarded as units subject to Darwinian evolution. It nonetheless remains true that the forces shaping natural selection among individuals involve all manner of biological interactions with other species: flowering plants with their pollinators, fruits with the animals that disperse their seeds, distasteful insects with the species that mimic their warning coloration. Therefore in a sense constellations of species can be viewed as evolving together within a conventional Darwinian framework. There is nothing of the crude notion of "group selection" in the recognition that evolution produces patterns at the level of ecological systems. Such patterns are anchored in the interplay of biological relations that act to confer specific advantages or disadvantages on individual organisms, a concept that has been captured memorably in the title of a book by G. Evelyn Hutchinson: *The Ecological Theater and the Evolutionary Play.*

How do evolutionary forces combine with the physical environment to shape a community of living things? Why, for example, does Britain have about 60 resident species of butterflies whereas New Guinea has close to 1,000? Going beyond the question of the total number of species, more detailed questions can be asked about the patterns of community organization. What accounts for the relative abundances of individuals among the various species in a given region? Why do some communities include several very common species along with a few rare ones whereas in other communities the individuals are distributed in roughly equal numbers among the species? Why do virtually all communities have many more species of little animals than of big ones? Why do food chains typically have only three or four levels (plant, herbivore, first carnivore, second carnivore) in spite of the fact that there is great variability in the amount of energy flow and in the physical details of the organisms involved?

The answers to such questions are not only interesting in themselves but also important for conservation and resource management. Notwithstanding a growing accumulation and synthesis of empirical evidence, however, there is at present no consensus on the answer to any of the above questions. I shall describe here a number of the empirical patterns that have emerged and review some speculations that have been put forward about the underlying causes.

Evidence for consistent patterns in the number of species associated with a given region comes from sources ranging from the grand sweep of the fossil record to controlled experiments on the arthropod fauna associated with individual mangrove trees. Drawing on that evidence, P. J. Darlington's *Zoogeography: The Geographical Distribution of Animals,* a classic study of the past and present whereabouts of the vertebrates, concludes: "Throughout the recorded history of vertebrates, whenever the record is good enough, the world as a whole and each main part of it has been inhabited by a vertebrate fauna which has been reasonably constant in size and adaptive structure. Neither the world nor any main part of it has been overfull of animals in one epoch and empty in the next, and no great ecological roles have been long unfilled. There have always been (except perhaps for very short periods of time) herbivores and carnivores, large and small forms, and a variety of different minor adaptations, all in reasonable proportion to each other. Existing faunas show the same balance. Every continent has a fauna reasonably proportionate to its area and climate, and each main fauna has a reasonable proportion of herbivores, carnivores, etc. This cannot be due to chance."

Similar evidence for a variety of fossil groups is presented elsewhere in this issue [see "The Evolution of Multicellular Plants and Animals," by James W. Valentine, page 66]. One particularly nice example is provided by a comparison of the number of families of land mammals in North America and South America before, during and after the formation of the Panama land bridge between the two continents in the Pleistocene epoch some two million years ago. Prior to the formation of this link the two continents had had no direct contact since the dawn of the age of mammals, and even before that the connection was circuitous, by way of Europe and Africa. Initially no families of land mammals were common to both continents, and given ecological roles were played by phylogenetically distinct actors in North America and South America [*see illustration on next page*].

IDEALIZED FOOD WEB, shown in the drawing on the opposite page, maps out what eats what in a willow forest in Canada. Such food webs can be roughly organized into a hierarchy consisting of a small number of "trophic levels" (in this case four), although usually not all the species present fit neatly into the classification scheme. Here the trophic levels are identified as primary producers (the willows themselves), herbivores (a variety of insects), first carnivores (spiders and frogs) and second carnivores (birds and snakes). The pathway from plants to snakes, however, can be traced either through two intermediate links (insects, frogs) or through three intermediate links (insects, spiders, frogs); similarly, the pathways from plants to birds can go either directly through one link (insects) or indirectly through two (insects, spiders).

A FULL CONTINENTAL COMPLEMENT of land mammals evolved independently in both North America and South America prior to the formation of the Panama land bridge during the Pleistocene epoch some two million years ago. As a result given ecological roles were played by phylogenetically distinct actors on each continent. For example, a number of such evolutionarily convergent types of mammals are identified by the lettered pairs. They include the shrew family in North America (*A*) and the caenolestine marsupial family in South America (*A′*), the North American wolf (*B*) and the corresponding South American marsupial carnivore (*B′*), the North American camel (*C*) and the South American camel-like litoptern (*C′*), the North American horse (*D*) and the South American horselike litoptern (*D′*), the North American rhinoceros (*E*) and the South American toxodont (*E′*), the North American chalicothere (*F*) and the South American homalodothere (*F′*), and the North American saber-toothed cat (*G*) and its South American marsupial counterpart (*G′*). After the present land connection was formed sometime in mid-Pleistocene, a great faunal exchange ensued between the two continents.

Sometime in the mid-Pleistocene the present land connection was formed, and there ensued a great faunal interchange between the two continents. The total number of families rose markedly during the interchange, but subsequent extinctions eventually caused the numbers of families on both continents to drop back to approximately the numerical level that existed before the faunal mixing. As the paleontologist George Gaylord Simpson has observed, these facts are consistent with "the idea that each continent was ecologically full of land mammals before interchange and that the number of about 25 families in North, about 30 in South America represents ecological saturation." Northern animals were more successful in the south than southern ones were in the north, probably because the North American species were the winnowed products of competition in the linked northern continents whereas South America had long been isolated.

On a much smaller scale such patterns of "island equilibrium" appear whether one counts the number of species of birds breeding on British offshore islands, the number of species of weevils on islands in the Pacific or (turning from real islands to virtual ones) the number of plant species in reserves of different sizes in Yorkshire. Underlying such studies is the idea that the number of species in an "island" community of this type is in dynamic equilibrium, with local extinctions being on the average compensated for by fresh immigration of the same species or other species.

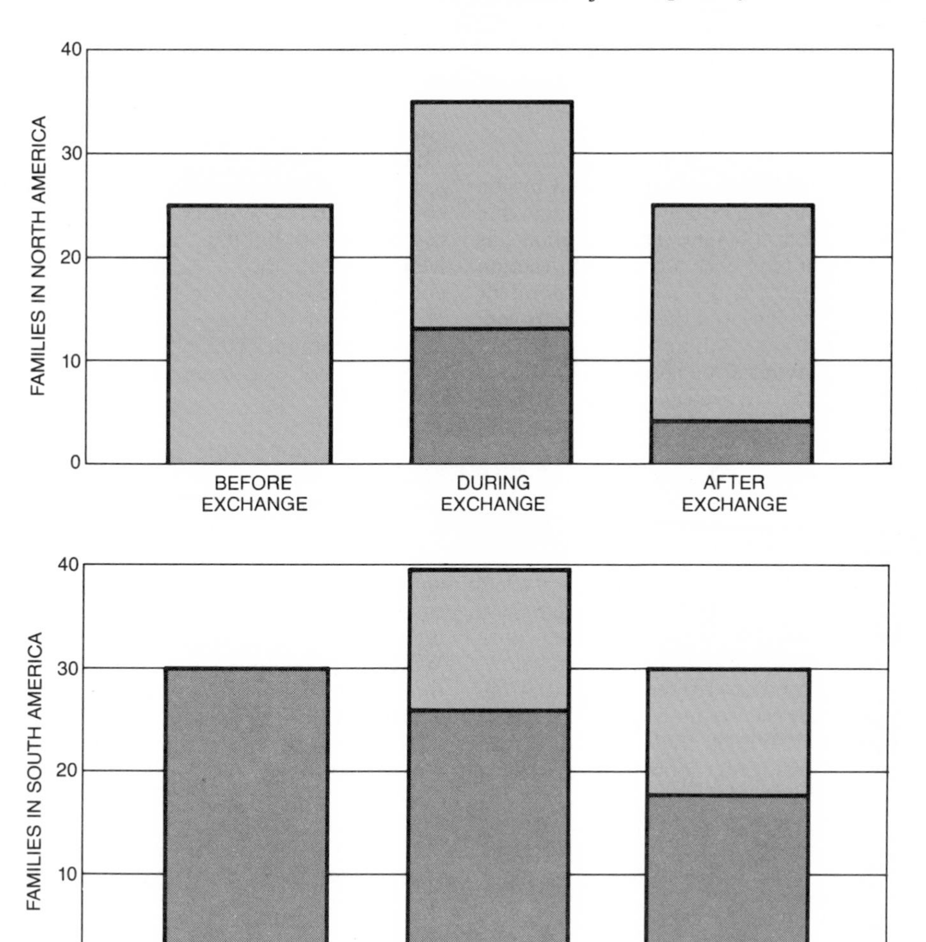

TOTAL NUMBERS OF FAMILIES of land mammals in North America (*top*) and South America (*bottom*) rose markedly during the faunal exchange between the two continents following the formation of the Panama land bridge, but subsequent extinctions eventually caused the totals to drop back to approximately the same numerical levels that existed earlier. Northern species (*color*) were generally more successful in the south than southern ones (*gray*) were in the north, as a result of greater geographical isolation of South America in earlier times.

The precise formulation of this idea is due to the late Robert H. MacArthur of Princeton University and Edward O. Wilson of Harvard University. Whether one is dealing with real or virtual islands, extinction rates are likely to be higher on smaller islands, so that the equilibrium number of species is likely to be lower on smaller islands. In New Guinea, for example, many species of birds are confined to higher elevations in the mountains; for them New Guinea itself therefore represents a kind of archipelago. More generally, the relation between the number of species and the area of an island (be it a real island, a hilltop or a wildlife preserve) in such an archipelago can be summed up with remarkable consistency in the rough rule that a tenfold increase in area results in a twofold increase in the number of species. Conversely, for an island of a given size the immigration rate is likely to diminish as the distance from the mainland species pool or from other islands increases. This effect has been documented for nonmarine lowland birds on the islands around New Guinea by Jared M. Diamond of the University of California at Los Angeles. Putting together the trends for extinction rates and for immigration rates, MacArthur and Wilson concluded that the equilibrium number of species will be highest on large, nearby islands and lowest on small, distant islands, provided that the islands are not grossly dissimilar in the habitats they provide.

There are essentially three ways to test these ideas about island biogeography. The first, described above, is to examine trends in species numbers on islands where the fauna appears to be in an equilibrium state. The second is to study the approach to equilibrium as empty islands are colonized. The best-known study of the latter type was provided by a "natural experiment": the recolonization of the volcanic island of Krakatoa after its biota had been destroyed by the explosion of 1883. Here the number of bird species returned in a fairly short time to the value appropriate to the island's area and isolation, whereas the number of plant species is still rising. The rate of approach to equilibrium obviously depends on the plant or animal group under consideration. Similar natural experiments are provided by the birds of Ritter Island and Long Island near New Guinea, whose faunas were destroyed by volcanic eruptions respectively in the late 19th and late 18th century, by the birds of seven coral islets in the same area where the tidal wave following the Ritter Island eruption destroyed the fauna in 1888, and by the initial colonization of a newly created volcanic island such as Surtsey, off Iceland.

In a series of analogous "artificial experiments" Daniel S. Simberloff of Florida State University has joined forces with Wilson to fumigate tiny mangrove islets off the Florida Keys and monitor their recolonization by various arthropods. These studies are particularly interesting because it was found that the total number of species tends to return to its original value, even though the actual lists of species on any one islet, before and after, are quite different. Furthermore, the more distant of these islets began and ended with fewer species of arthropods, as had been predicted by the equilibrium theory.

The third way to test ideas about island biogeography is illustrated by the case of an oversaturated island relaxing toward equilibrium. This situation has practical relevance when some fraction of a habitat is set aside as a floral or faunal reserve and the rest is destroyed; such a reserve will at first be supersaturated, having more species than are appropriate to its area at equilibrium. Natural experiments of this kind are provided by the land-bridge islands. In the last ice age these island were attached to the mainland or to larger islands, and they shared the continental biota. At the end of the ice age, about 10,000 years ago, the release of water stored in glaciers raised the sea level and created these islands, with the result that their continental complement of species has slowly decreased toward the equilibrium value appropriate to an island of their area today. Examples that have been analyzed by biogeographers include Britain (off continental Europe), Aru and other islands (off New Guinea), Trinidad (off South America) and Borneo and Japan (off the mainland of Asia).

A variation on this theme is provided by 17 mountain ranges that rise to elevations of more than 10,000 feet out of the Great Basin desert of the western U.S. The boreal habitats at their summit are now "mountain islands," but they were connected to one another and to the ancestral boreal habitats of the Rocky Mountains and the Sierra Nevada in the cooler periods of the Pleistocene. James H. Brown of the University of Arizona has documented the regular patterns of differential extinction that have occurred as the supersaturated faunas of 13 species of small flightless mammals have relaxed toward the smaller number of species that are appropriate to particular mountaintops [*see illustration on page 87*].

These examples and others point to an underlying community structure, an equilibrium number of species, which on the average is steady and predictable. Which species are in fact present, on the other hand, can be quite unpredictable and can depend on the whims of evolutionary history or the vagaries of the environment. Anecdotal instances abound. In both Old World and New World environments woodpeckers occupy a distinctive niche, taking insects from under the bark of trees. Woodpeckers have not reached the Galápagos Islands, however, where their role is filled by a tool-using finch that probes for the insects with a cactus spine; nor have they spread to Hawaii, where a bird of the honeycreeper family has evolved a woodpeckerlike beak; nor to New Guinea, where various birds and the striped opossum have expanded into their niche; nor to Madagascar, where the woodpecker's life-style is imitated by the aye-aye, a primate whose third finger is grotesquely elongated to enable it to extract insect larvae from trees.

In the African and South American tropical forest, monkeys are conspicuous among the fruit eaters, but monkeys have not crossed from the Asian biogeographical realm into New Guinea and as a result New Guinea has a greater diversity of fruit-eating birds. This is all evocative of the perceptions of the evolutionary process set forth in Jacques Monod's book *Chance and Necessity,* albeit here at the level of ecosystems. In Monod's terms the patterns of community organization are predictable and therefore necessary, but the species that happen to play a designated ecological role at a particular place and time are subject to historical accident, or chance.

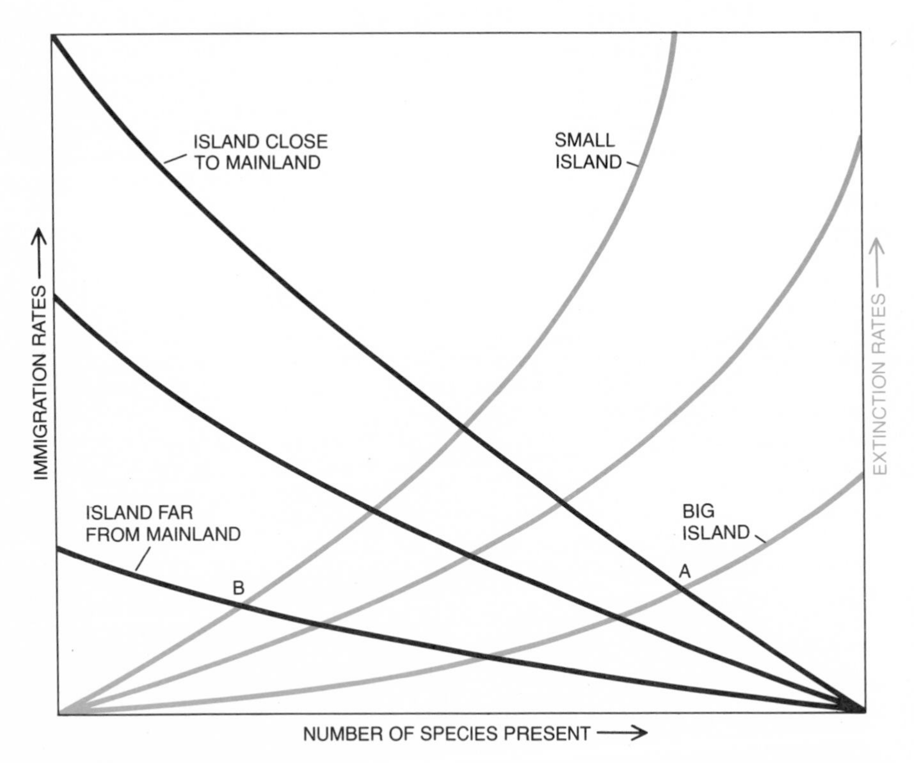

DYNAMIC-EQUILIBRIUM MODEL of island biogeography, formulated by the late Robert H. MacArthur and Edward O. Wilson, holds that in both real and virtual islands local extinctions will on the average be compensated for by fresh immigration of the same species or other species. Extinction rates (*colored curves*) will tend to rise as the total number of species on the island rises, and this effect will be more pronounced on small islands than on big ones. Conversely, the rate of immigration of new species (*black curves*) will tend to fall as the number of species rises, with the immigration rate being lower for islands far from the main species pool than for nearby ones. It follows that the equilibrium number of species present in a given area will be highest on the large, close islands (*point A*) and lowest on small, distant islands (*point B*).

It is easier to demonstrate the existence of these patterns than to explain them. Current work builds on the pioneer efforts of Charles Elton, Hutchinson and MacArthur, but it remains more a list of ideas than a catalogue of answers. Suppose one begins by focusing attention on the individual populations that constitute a multispecies community. An immediate complication is the variety of factors that can bear on whether or not a particular population is likely to persist. At one extreme there are species whose mortality patterns are determined almost entirely by the vicissitudes of an environment that to them is unpredictable, transient or patchy. For them the evolutionary pressures are to produce many young, to invest little in parental care (since it has scant effect on survival probability) and to spread the risks by high dispersal ability. Such "boom and bust" species are the pioneers in the world of plants and animals. At the other extreme are species whose mortality is predominantly influenced by interactions with their own species and other species. For them the evolutionary pressures are to be a good competitor and to have few offspring but to invest more time and energy in raising them. These are of course the two extremes of a continuum, and most species occupy some intermediate position.

Practical consequences attend this recognition of the variety of possible life-styles. For example, as has been stressed by T. R. E. Southwood of the Imperial College of Science and Technology in London, there can be no one "royal road" to controlling insect pests. For pests such as desert locusts or fruit flies, whose natural history is one of vagarious outbreak and crash, one cannot appeal to "the balance of nature," because there is none. Pesticides, intelligently applied, will continue to be the most effective technique for coexisting with these inherently booming populations. Other pests, such as the codling moth, inflict economic damage as they persist at low and steady values. Here control by biological techniques such as natural enemies, release of sterile males or the use of pheromones (behaviorally active natural substances such as sex attractants) may prevail. The organisms associated with human diseases provide another class of examples. At one end of the spectrum are diseases such as smallpox, influenza or measles, which are epidemic, short-lived in any one individual and highly transmissible by direct contact. At the other end are infections such as malaria or schistosomiasis, which are endemic, long-lived and with recondite but stable transmission cycles involving all manner of intermediate vectors.

Suppose next one forgets these complications and concentrates on communities in which biological interactions are all-important in determining the number of species that coexist. Among species that compete for a resource the key question then is how similar two species can be and yet persist together. In other words, what are the limits to similarity among coexisting competitors? This question, first put in precise form by Hutchinson and MacArthur, has been pursued in theoretical models and in the real world. In both cases simple situations need to be considered initially, lest simultaneous competition for several resources (food, foraging places, nest sites and so on) lead to complications that are not easily unraveled. For species that compete primarily along a single resource axis (for example, sorting themselves out solely by the size of the food items they select) some field observations and several different lines of theoretical reasoning suggest that in the utilization of the resource (the food size selected) the average difference between species must be greater than, or approximately the same as, the spread within either species, if they are to coexist. These are as yet tentative and preliminary answers to the question of the limits to similarity.

Many observers feel that competition cannot be of much importance in real communities, because direct evidence of it—blood on the ground, as it were—is rarely seen. To this argument Diamond has replied with an analogy: the Hertz and Avis attendants, each with their distinctive species colors, are rarely seen locked in physical combat at airports, yet their indirect competition for car renters is nonetheless real.

Other kinds of biological interaction can profoundly modify the effects of competition. Thus predators can promote the coexistence of prey species. This is obviously the case if the predators switch their attention to concentrate disproportionately on the prey species that are most abundant at any one time. It can also, however, happen with predators that take prey species indiscriminately, provided that among the prey species the inferior competitors have higher rates of population growth. Mutualistic interactions, such as those between flowering plants and their pollinators, or between ants and the plants that provide them with special houses in return for defense against herbivorous insects, can also enrich community structure and species number. In effect mutualistic interactions can create new resources.

As if all these complications were not enough, there are fascinating but troublesome problems associated with the simplest models for the dynamical behavior of single populations. Just as the mathematical equations that describe the flow of fluids can exhibit behavior

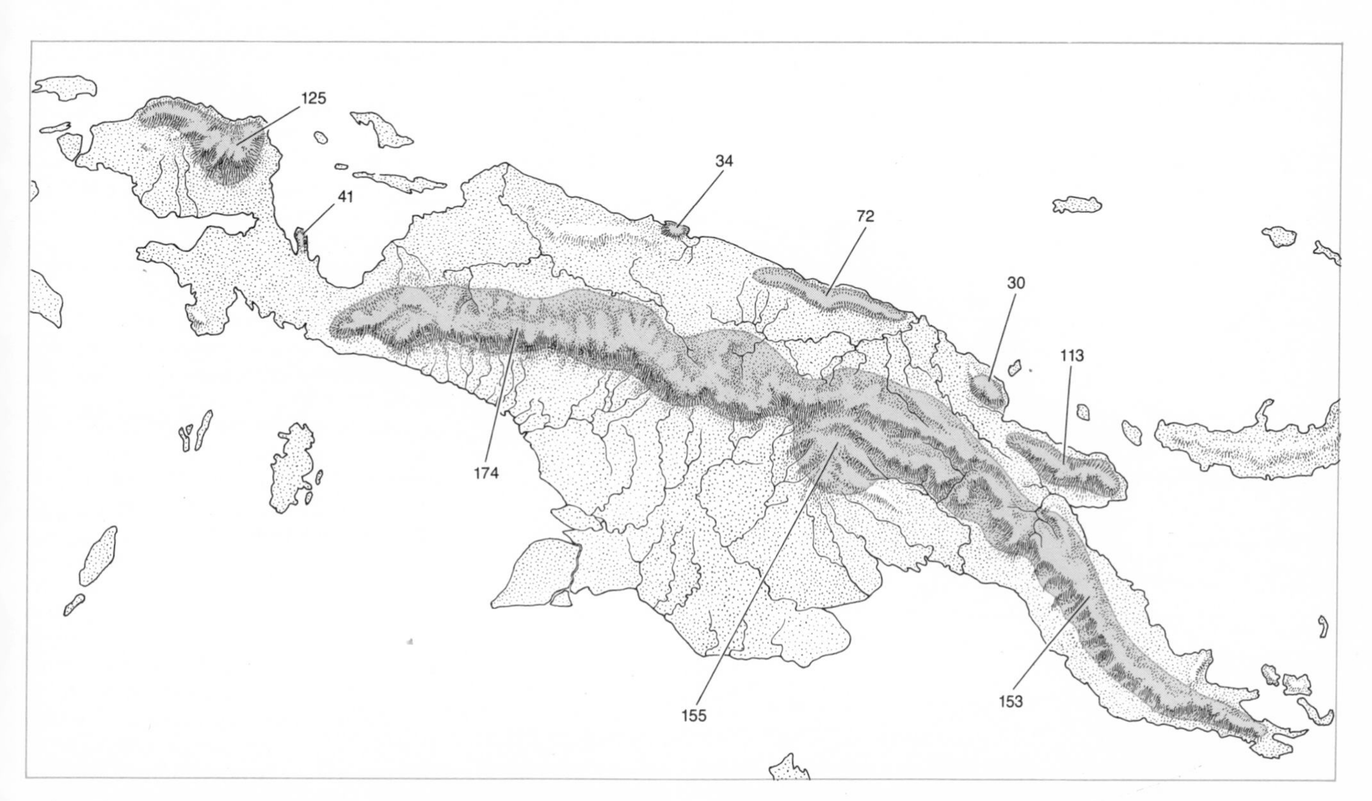

MOUNTAINS OF NEW GUINEA (*color*) represent a kind of island archipelago for the species of birds that are confined to higher elevations. For such species New Guinea in effect consists of a large central island habitat (or mountain range) and about six smaller island habitats along the northern coast, separated from one another by a "sea" of intervening lowlands. The numbers on the map indicate the number of highland bird species counted on each small mountain range and at three different locations on the central range. In keeping with the dynamic-equilibrium model of island biogeography the larger virtual islands generally have more species. This area effect explains most of the variation in the observed number of species; the residual variation is largely correlated with variation in altitude.

corresponding to steady flow, oscillatory flow or turbulence, so in a similar (but more transparent) way nonlinear equations for population growth can give steady solutions, population cycles or chaotic population behavior. This understanding, which has recently emerged from the work of James A. Yorke of the University of Maryland, George F. Oster of the University of California at Berkeley, myself and others, helps to explain why some populations are steady and others exhibit cycles. On the other hand, it raises the unwelcome prospect that a realistic understanding of multispecies systems may bog down in the same kind of mathematical difficulties afflicting the study of turbulent fluids.

In short, ecologists are a long way from explaining why, say, there are about 600, rather than 60 or 6,000, species of birds in North America. My personal belief is that such limits to the total number of species are set mainly by limiting similarity among competitors and are often modified by predator-prey relations, by mutualistic effects and by the existence of fugitive, boom-and-bust species that make their living on the run. To these effects must be added the history of the region: the older and more climatically stable the habitat, the more intricate the species relations generated by evolution and hence the larger the total number of species. The next decade may see these issues clarified, but the complexities inherent in nonlinear systems may still block many empirical and theoretical advances.

Let us now turn from the total number of species in an ecological system to delve into some of the detailed patterns of the relations among species. One such pattern involves the relative abundance of the various species. A typical tendency is for early successional communities, frequently disturbed ones or those in harsh environments to have one or two dominant species, accounting for most of the individuals present. Later stages of succession tend not only to have a larger total number of species but also to have a more "middle class" distribution of individuals among species, with many roughly equally abundant ones. For example, the illustration on page 130 depicts the changing patterns of the relative abundances of plant species as succession unfolds in an abandoned field; the features shown are characteristic of the successional process. A possible reason is that in early succession or in harsh environments the ecological factors bearing on community organization are comparatively few (or are no better than random in determining which species colonize an empty patch first), so that the most successful or the first-arriving species preempt most of the available "niche space." Conversely, in later and more biologically crowded successional stages the interplay of a large number of ecological factors leads to a comparatively uniform distribution, as success takes many forms.

If a mature ecosystem is seriously disturbed, the distribution of relative abundances tends to revert to the level characteristic of early succession, dominated by a comparatively few species. Ruth Patrick of the Academy of Natural Sciences in Philadelphia has repeatedly demonstrated the fact in studies of the communities of diatoms in streams and lakes subject to disturbance ranging from "enrichment" by waste heat, sewage, excess nitrogen or excess phosphorus to pollution with various toxic substances. Among other examples are the experimental grass plots at the Rothamsted Experimental Station in England. These plots were set aside in the middle of the 19th century. Each was subjected to some specified treatment, such as the withholding or overapplication of certain fertilizers. The resulting trends in the relative abundances of the grass species present are similar to those obtained by Patrick for her diatoms, and they look like a successional sequence run backward. A plausible explanation of these trends is that such disturbances, whether they are toxic or enriching, distort community organization by emphasizing one ecological factor at the expense of all other factors, leading to strong dominance by the handful of species that cope best with the single factor.

A conspicuous feature of most communities is the presence of many more species of small creatures than of big ones. There are, for example, roughly as many species of beetles in Britain as there are species of mammals in the entire world. An understanding of the way ecological systems are organized will need to take account of these trends in the number of species of organisms in different size classes.

As a first step I have made a very rough compilation of the numbers of

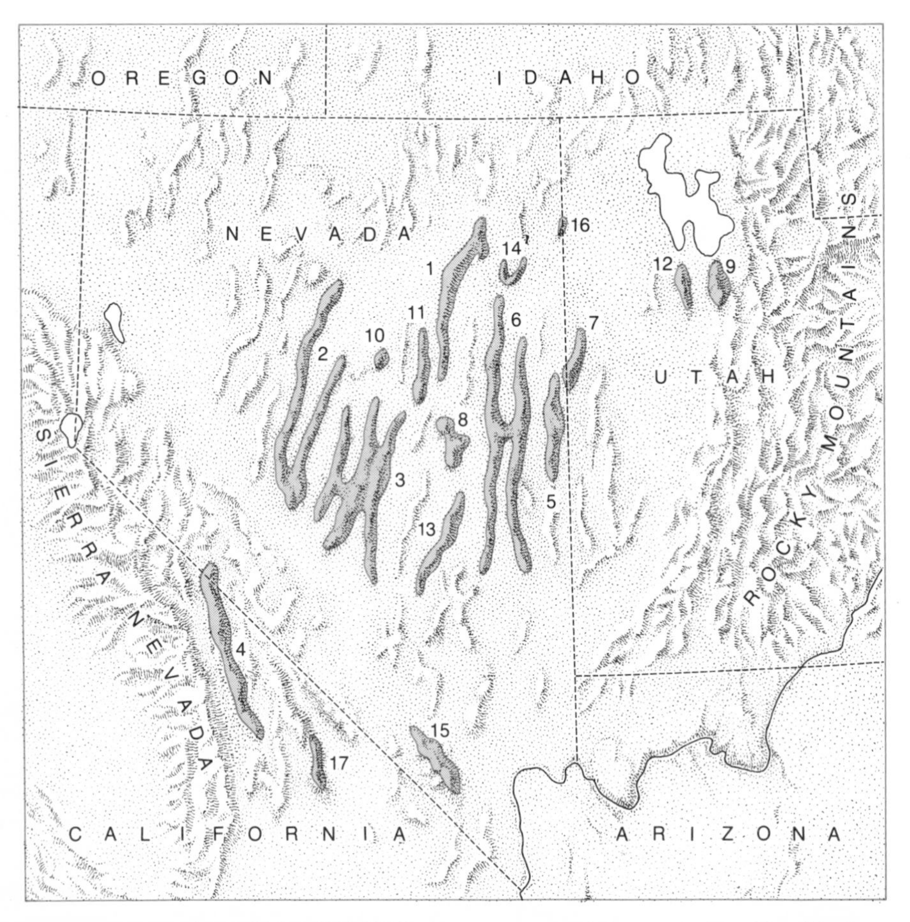

SPECIAL CASE IN BIOGEOGRAPHY of an oversaturated "mountain island" archipelago that is relaxing toward equilibrium is represented on this map by a group of 17 mountain ranges higher than 10,000 feet in the Great Basin desert region between the Sierra Nevada and the Rocky Mountains. According to James H. Brown of the University of Arizona, the climatically isolated boreal habitats now found on these summits were connected to one another and to the ancestral boreal habitats in the neighboring mountain ranges during the cooler Pleistocene epoch, when they were colonized by a group of small animals. For each of the virtual islands Brown has documented pattern of extinction of 13 species (*see illustration on opposite page*).

MOUNTAIN RANGE

Species	Body weight (grams)	1. Ruby	2. Tioyabe	3. Toquima, Monitor	4. White, Inyo	5. Snake	6. Schell Creek, Egan	7. Deep Creek	8. White Pine	9. Oquirrh	10. Roberts Creek	11. Diamond	12. Stansbury	13. Grant	14. Spruce	15. Spring	16. Pilot	17. Panamint
BUSHYTAIL WOOD RAT	317	■	■	■	■	■	■	■	■			■	■		■	■	■	■
UINTA CHIPMUNK	57	■	■	■	■	■	■	■	■	■	■	■		■	■	■		
GOLDEN-MANTLED GROUND SQUIRREL	147	■	■	■	■	■	■	■	■			■		■	■	■	■	
PRAIRIE VOLE	47	■	■	■	■	■	■	■	■	■	■		■	■				
YELLOW-BELLY MARMOT	3,000	■	■	■	■	■	■	■	■				■					
NORTHERN POCKET GOPHER	102	■	■	■	■		■		■	■		■						
VAGRANT SHREW	7	■	■			■	■	■		■								
NORTHERN WATER SHREW	14	■	■	■	■	■					■							
WESTERN JUMPING MOUSE	33	■	■							■	■							
PIKA	121	■	■	■	■													
SHORTTAIL WEASEL	58		■		■	■												
BELDING GROUND SQUIRREL	382	■	■	■														
WHITETAIL JACKRABBIT	2,500	■																

PATTERNS OF EXTINCTION of 13 small mammalian species result from the relaxation of the supersaturated faunas of the 17 Great Basin mountain habitats toward the smaller species numbers appropriate to particular mountaintops. The mountain ranges are named at the top, along with the numbers that indicate their location on the map on the opposite page. Brown refers to this case as an example of "nonequilibrium insular biogeography," since "the mammalian faunas of the mountaintops are true relicts and do not represent equilibria between rates of colonization and extinction." The colored cells denote the locations of the surviving species. (The 13 species shown in the illustration are not depicted to the same scale; their relative sizes are indicated by giving a typical body weight for each species.)

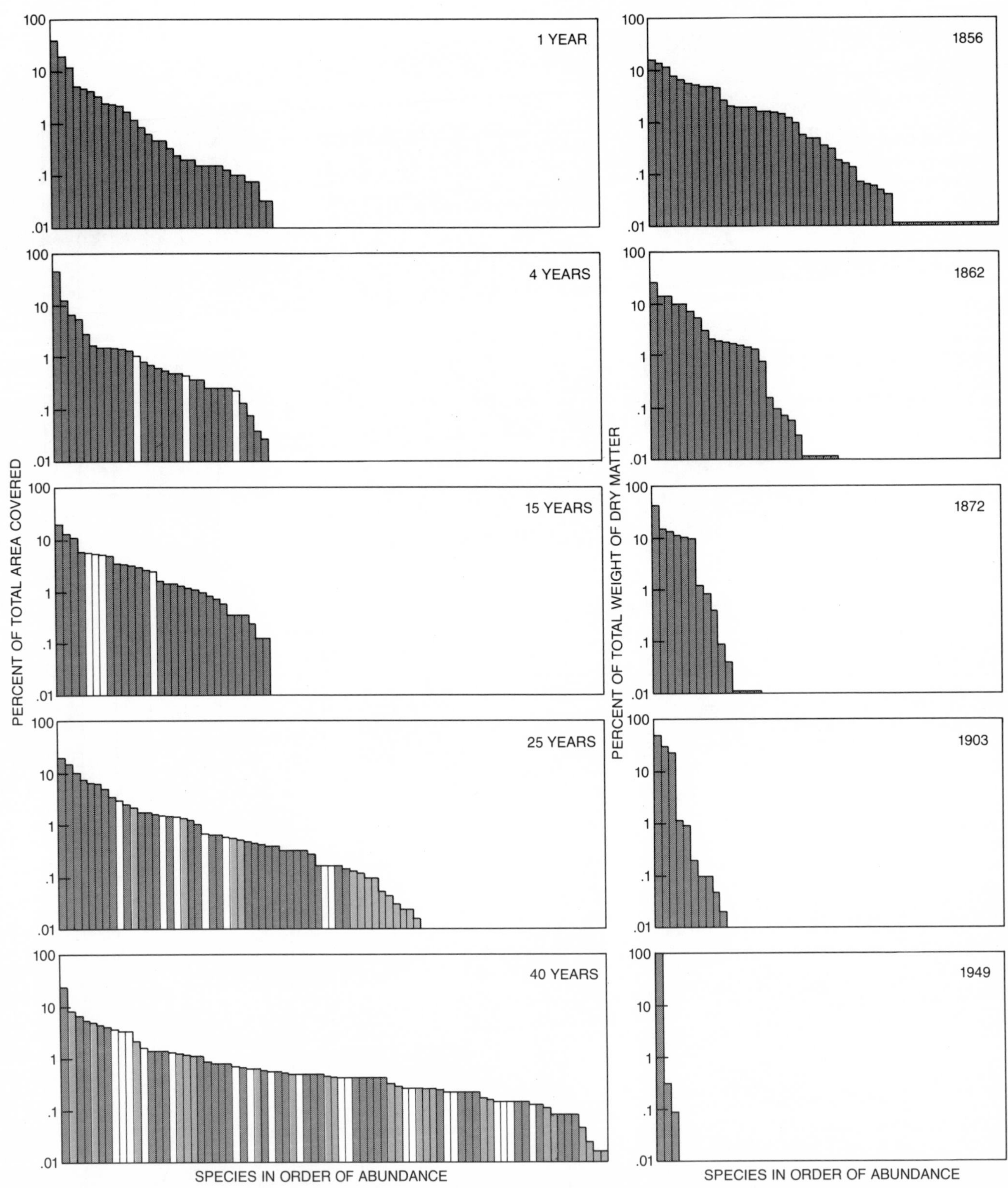

PATTERNS OF RELATIVE ABUNDANCE of various species in two different kinds of ecological community are contrasted. The series of graphs at the left is typical of an early successional community, in this case an abandoned agricultural field in southern Illinois that has been studied by Fakri A. Bazzaz of the University of Illinois. The plant species present in the field were counted at five stages, ranging from one year to 40 years after the fields were abandoned. The observed relative-abundance patterns are expressed in terms of the percentage each species contributes to the total area covered by all the species in the community, plotted against the rank of the species, ordered from the most abundant species (*left*) to the least abundant (*right*). The gray bars are herbs; the white bars are shrubs; the colored bars are trees. As in most such communities the overall trend is from dominance to diversity, that is, from a situation in which a few species are dominant to one in which there is a large "middle class" distribution of individuals among many species. The series of graphs at the right is typical of a mature ecosystem that has been severely disturbed, in this case an experimental grass plot at the Rothamsted Experimental Station in England to which nitrogen fertilizer has been continuously applied since 1856. An analysis conducted by W. E. Brenchley, K. Warington, R. A. Kempton and R. L. Taylor of the Rothamsted Station shows that the resulting relative-abundance patterns, measured at sporadic intervals over the following century, look like a normal early-successional sequence that has been run backward. Presumably the disturbance had the effect of distorting the organization of this pasture community by emphasizing one ecological factor at the expense of all the others, leading to the strong dominance of a single grass out of the three surviving species of grasses.

species of terrestrial animals as a function of the physical size (specifically the characteristic length) of their constituent individuals [*see illustration on page 90*]. There are many difficulties in this exercise. One is that the systematics of small arthropods and other invertebrates are in most instances in a rudimentary state. As Ernst Mayr has put it: "We must take it for granted that a large part of the mite fauna of the world will remain unsampled, unnamed and unclassified for decades to come." This could easily mean that the classes of animals smaller than a centimeter in length are underestimated in the illustration by a factor of two or more. Another difficulty, which might account for the decrease in numbers of species of very small size, is that conventional taxonomic concepts have dubious validity once one goes below the one-millimeter size class.

Setting aside these caveats, the overall trend for organisms ranging in size from about a centimeter to a few meters is that a threefold increase in length corresponds roughly to a tenfold decrease in the number of species. Part of the explanation of this trend is clearly that small animals can subdivide the habitat more than large ones can; a species of small plant may be merely one among many in the diet of a large herbivore, and yet it may provide a rich variety of different niches wherein several species of small arthropods can coexist.

Within a given community the biological relations among species can be depicted as a food web in which the links between pairs of species map out what eats what. More abstractly, such webs trace the paths, or food chains, by which energy flows through the ecosystem. To a very rough first approximation, food webs can be organized into a hierarchy of "trophic levels" from primary producers (green plants) through herbivores to various categories of carnivores. In the idealized food web shown on page 80, for example, four trophic levels can be roughly identified, although here, as in most real food webs, some of the relations do not fit tidily into the classification scheme.

The food web depicted in this example is typical of ecological systems in the real world, where food chains are characteristically short, rarely consisting of more than four or five trophic levels. Emphasizing this point, Stuart Pimm of Texas Tech University has analyzed data compiled by Joel Cohen of Rockefeller University for 19 food webs, which include terrestrial, freshwater and marine examples. These webs contain a total of 102 top predators (animals themselves free from predation). Pimm and Cohen have independently traced out all the food chains connecting top predators to basal species (plants, detritus or arthropods that fall into freshwater systems). The average number of trophic levels is fairly consistently about three; for only one of the 102 top predators can a food chain involving more than six species (five links) be found.

Such steady patterns in the number of trophic levels are in pronounced contrast to the great variability in the amount of energy flowing through different ecological systems. Primary productivity varies over three to four orders of magnitude in both terrestrial and aquatic ecosystems, and the productivity of fish and terrestrial animal populations varies over five or more orders of magnitude. There are further variabilities in the efficiency of energy transfer from one level to the next, with such efficiencies typically being much low-

SURROGATE WOODPECKERS, animals that occupy the woodpecker's ecological niche in places where woodpeckers are absent, are depicted in this idealized scene. The species shown include a Galápagos finch, which uses a cactus spine to probe for insects under the bark of trees (*a*), a Hawaiian bird of the honeycreeper family, which has evolved a woodpeckerlike beak (*b*), a striped opossum from New Guinea (*c*) and the aye-aye, a primate from Madagascar (*d*). Both mammals have evolved long fingers that enable them to extract larvae from trees.

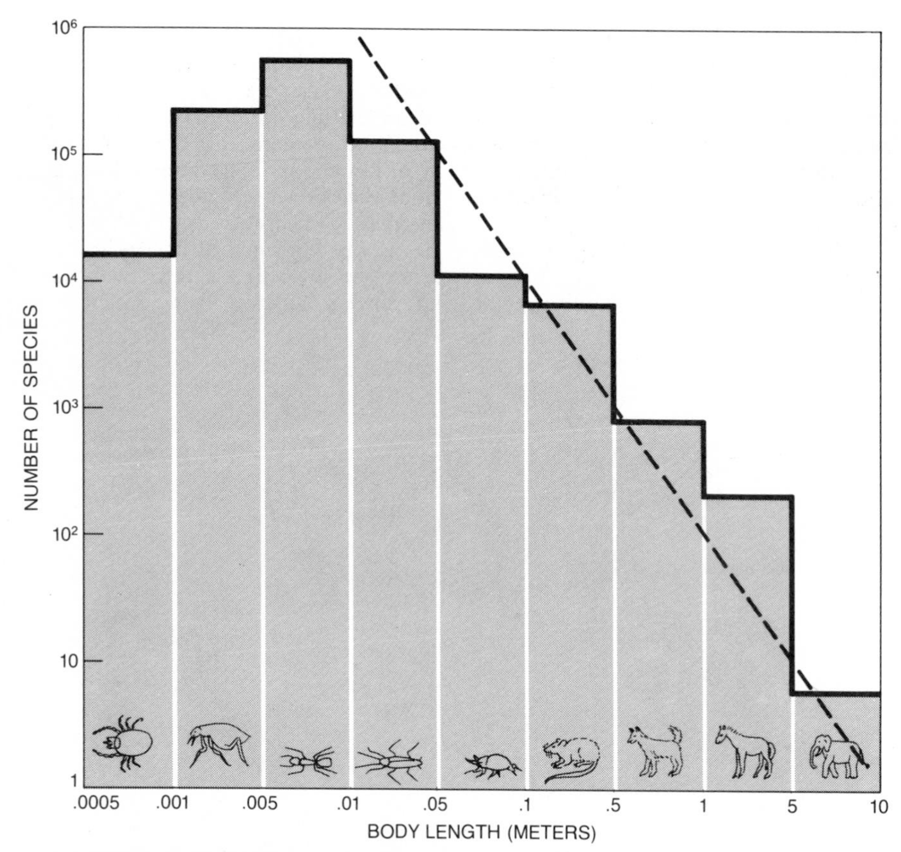

PREVALENCE OF SMALL ANIMALS over big ones is evident in this rough compilation of the numbers of species of terrestrial animals as a function of the characteristic body length of their constituent individuals. Disregarding species smaller than a centimeter in length, which present difficult problems of classification, the overall trend (*broken line*) is that a threefold increase in length corresponds approximately to a tenfold decrease in the number of species. Some representative animals are shown in the illustration, all of them drawn to the same scale.

er for warm-blooded animals than for cold-blooded ones.

The conventional explanation of the small number of trophic levels is that they are determined by energy flow; if only 10 percent of the energy entering one level is effectively transferable to the level above it, the number of levels is clearly limited. As Pimm and John Lawton of the University of York have recently observed, however, this explanation is not easily reconciled with the observation that the number of trophic levels is essentially independent of enormous variations in the amount of energy flow and in the transfer efficiencies. In Pimm and Lawton's words, "food chains are not noticeably shorter in barren Arctic and Antarctic terrestrial ecosystems compared with a productive tropical savanna or the fish guilds of a tropical coral reef."

Pimm and Lawton alternatively suggest that the explanation may lie in the dynamics of the various populations in the community. They argue, with the help of mathematical models, that long food chains may result in population fluctuations so severe that it is hard for top predators to persist. This notion is itself debatable, on the grounds that it is derived from rather special assumptions about community dynamics. The suggestion is, incidentally, representative of several recent studies that seek to understand the structure of ecological systems in terms of the dynamic properties of the interacting species. Another example is the set of studies seeking to elucidate the relation between the stability of an ecosystem (its ability to withstand disturbance) and the complexity of its food-web structure (the number of species and the number of connections among them).

The major theme of this article is reflected accurately in the discussion of the length of food chains. Here, as elsewhere, the empirical patterns are important, widespread and abundantly documented, but they lack a convincing explanation. So it is too for the patterns that can be discerned in the total number of species, in their relative abundances and in the distribution of species among size classes: in place of fundamental explanations there are only lists of possibilities to be explored. The task of understanding how ecological systems work is in the middle of its own successional process.

VII

The Evolution of Behavior

The Evolution of Behavior

BY JOHN MAYNARD SMITH

Here one of the key questions has to do with altruism: How is it that natural selection can favor patterns of behavior that apparently do not favor the survival of the individual?

Most species of gulls signal appeasement in fighting by turning their head sharply away from their opponent. This clearly identifiable display is called head flagging. Young gulls do not signal in this way; if they are threatened, they run to cover. One gull species, however, has proved to be an exception to the rule. Chicks of the ledge-nesting kittiwake species do employ the head-flagging display when they are frightened. Their anomalous behavior is the result of the interplay between innate behavior patterns and environmental forces. Unlike other gull species, which live on beaches, the kittiwake perches on tiny ledges of steep cliffs where there is no cover to which the chicks can run if they are threatened. The kittiwake species has responded to environmental pressures by accelerating the development of a standard motor pattern of adult gulls.

This explanation reflects a major change in the understanding of animal behavior. Formerly animal behavior was thought to consist of simple responses, some of them innate and some of them learned, to incoming stimuli. Complex behavior, if it was considered at all, was assumed to be the result of complex stimuli. Over the past 60 years, however, a group of ethologists, notably Konrad Z. Lorenz, Nikolaas Tinbergen and Karl von Frisch, have established a new view of animal behavior. They have shown that the animal brain possesses certain specific competences, that animals have an innate capacity for performing complex acts in response to simple stimuli.

The discovery that certain behavior patterns are inherited was an important contribution to the study of evolution. Genetically determined responses must be subject to the pressures of natural selection. Hence innate behavior must evolve. The ethologists were able to show how a motor pattern employed in a noncommunicatory context such as feeding could evolve into a ritualized form employed as a signal in, say, courtship. Differentiation in innate behavior patterns could be traced, as they were in the kittiwake, to selection pressures arising from the environment.

The concept of the evolution of behavior solved some problems and raised others. Since the time of Darwin morphological structures have been used to identify phylogenetic relations. For example, the similarity between a man's arm and a bat's wing is taken as evidence of their common origin. Lorenz pointed out that similarities in behavior patterns can also serve in reconstructing evolutionary history.

It is not always clear, however, how certain types of innate behavior evolved through natural selection. In its modern form the Darwinian interpretation of evolution asserts that (1) evolution consists in changes in the frequency of appearance of different genes in populations and (2) the frequency of the appearance of a particular gene can only increase if the gene increases the "Darwinian fitness" (the expected number of surviving offspring) of its possessors. There are many instances of animal behavior patterns that do not seem to contribute to the survival of the individual displaying that behavior. The classic example is the behavior of the worker bee: this insect will sting an intruder and thereby kill itself in defense of the hive. The problem is evident: How can a gene that makes suicide more likely become established?

The concern over this type and other types of apparently anomalous behavior led to the development of a new phase in the study of the evolution of behavior: a marriage of ethology and population genetics. From this perspective it has been possible to explain how natural selection operates to bring about the evolution of many of the most perplexing examples of animal behavior. In this article I shall discuss the progress that has been made in understanding the evolution of two important types of behavior: cooperative or altruistic behavior such as that of the worker bee and ritualized behavior in animal contests. I shall begin with the first type of behavior, one of the initial problems to which the new discipline was successfully applied.

For a long time many biologists, particularly those unfamiliar with genetics, explained the evolution of behavior such as the altruism of the worker bee by saying that this type of behavior contributed to the "good of the species." A behavior pattern that promoted the survival of a species would, they believed, be favored by natural selection even if it reduced the Darwinian fitness of the individual displaying it. There is an obvious problem with that explanation: if a gene increases the fitness of an individual, then it will be established in a species even if it reduces the long-term survival of the species.

Darwin and later the founders of population genetics, R. A. Fisher, Sewall Wright and J. B. S. Haldane, were aware of the problem and even came close to solving it. The current understanding of the evolution of altruistic behavior, however, is due to the work of W. D. Hamilton of the Imperial College of Science and Technology in London. Hamilton presented his thesis in two papers published in 1964. To understand Hamilton's argument, consider the fact that a parent may risk its life in defense of its offspring, say by feigning injury to distract a predator. In this way the parent may increase its own Darwinian fitness. Although it is possible that both the parent and the offspring will be killed, it is more likely that both the parent and the offspring will survive. In the latter case the parent's Darwinian fitness will be greater after the altruistic act than it would have been if the parent had left its offspring to the predator. The genes associated with the altruistic act (in this instance feigning injury) may be present in the offspring, so that their frequency may be increased. Hence natural selection favors parental altruism, that is, it is through parental altruism that the par-

DIFFERENTIATION in innate behavior patterns is the result of selection pressures arising from the environment. For example, adult gulls signal appeasement in fighting with a standard head-flagging display, turning their head sharply away from an opponent (*top*). Most young gulls do not employ the display; if they are threatened, they run to cover. Chicks of the ledge-nesting kittiwake species, however, do employ the head-flagging display when they are threatened (*bottom*). Unlike other gull species, which live on beaches, the kittiwake lives on tiny ledges of steep cliffs where there is no cover to which the chicks can run. Early development of the head-flagging behavior pattern contributes to the survival of the gulls. Hence natural selection favors the evolution of the kittiwake's anomalous behavior.

ent's behavioral characteristic is established in future generations. Hamilton realized that this analysis of parental altruism could also be applied to explain acts increasing the chances of survival of relatives other than children, for example siblings or even cousins. It was this basic perception that was the key to understanding the evolution of a wide range of animal behavior.

Consider two individuals, a "donor" and a "recipient." The donor performs an act that reduces its own Darwinian fitness, or expected number of surviving offspring, by a cost C but increases the recipient's fitness by a benefit B. Suppose that there is a pair of allelic, or alternative, genes A and a and that the presence of A makes an individual more likely to perform the act. Hamilton showed that the change in the frequency of gene A in the population after the act depends on the coefficient of relationship r between the donor and the recipient, that is, on the average fraction of genes of common descent in individuals with the genetic relationship of the donor and the recipient. More precisely, he showed that (with certain approximations) the frequency of gene A will increase because of the donor's act if the coefficient of relationship r is greater than C/B.

For example, if an individual has two sets of genes, one from a father with two sets and one from a mother with two sets, then on the average the probability is 1/2 that any particular gene in the individual is present in a full sibling [*see illustration on this page*]. Hence the coefficient of relationship between the individual and its full sibling is 1/2. Therefore, according to Hamilton's argument, if a gene in the individual causes it to sacrifice its life to save the lives of more than two siblings, then the number of replicas of the gene present after the sacrifice is made is greater than the number present would be if the sacrifice had not been made. The sacrifice is selectively advantageous. (In this instance the cost C is equal to 1 and the benefit B is equal to more than 2, so that the coefficient of relationship 1/2 is indeed greater than C/B.)

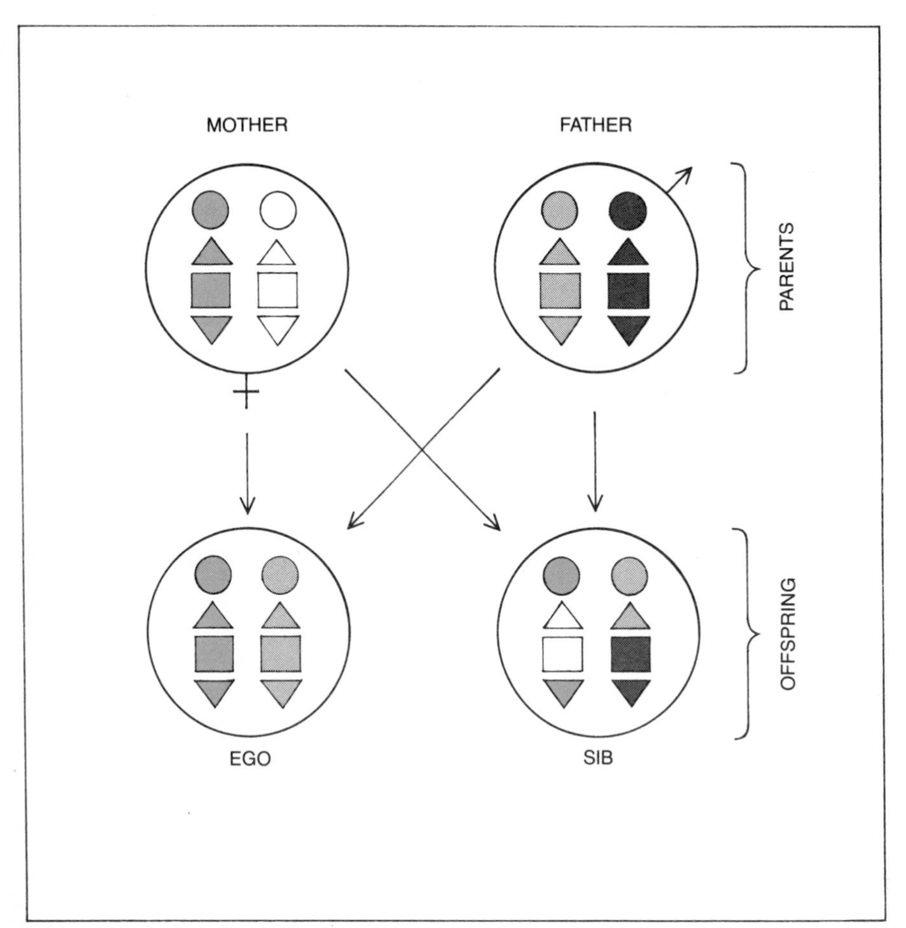

ALTRUISTIC ACTS do not appear to contribute to the survival of the animals performing them, but their evolution can be understood by examining the genetic relationship between the performer of an act and the beneficiary of the act. The genetic relationship between any two individuals can be expressed by a coefficient of relationship that is defined as the average fraction in the individuals of shared genes, or genes of common descent. For example, this illustration shows two parents, each with two sets of four genes, and two offspring: "Ego" and a full sibling, "Sib." The four pairs of alleles (alternative forms of the same gene) in each individual are represented by four different shapes. The color of the genes in the offspring indicates the manner in which parental alleles have been reassorted. Ego and Sib each have two sets of inherited genes, one from their mother and one from their father, so that on the average the probability is 1/2 that a gene present in Ego is also present in Sib. Hence the coefficient of relationship between the two siblings is 1/2. Modern evolutionary theory states that evolution consists in changes in the frequency of the appearance of various genes in a population and that a gene can only increase in frequency if it increases the Darwinian fitness, or expected number of surviving offspring, of its possessor. W. D. Hamilton of the Imperial College of Science and Technology in London showed that (with certain approximations) the gene associated with an altruistic act will increase in frequency because of the act only if the coefficient of relationship between the performer and the beneficiary is greater than *C/B*, where *C* is the cost (in Darwinian fitness) of the act to performer and *B* is benefit (in Darwinian fitness) of act to beneficiary.

Hamilton's work predicts that altruistic and cooperative behavior will be found more frequently in interactions of related individuals than in interactions of unrelated individuals. Observation certainly confirms this prediction. In fact, as Hamilton points out, the highest degree of cooperation is displayed by colonies of genetically identical cells such as the cells that make up the human body. It is important to note that these concepts apply to organisms incapable of recognizing degrees of relationship. In species that usually live in family groups a gene causing an individual to act altruistically toward members of its community will increase in frequency even if the individuals carrying it cannot recognize family members. The warning signals given by birds and mammals (such as a rabbit's thumping the ground with its hind legs) exemplify this type of altruistic behavior.

One of the best illustrations of the same type of altruism is the behavior of a viruslike self-replicating particle called a plasmid, which lives as a parasite in bacteria. On occasion a plasmid manufactures a toxin that kills the host bacterium and probably the plasmid as well. When the host dies, the toxin is released, but it kills only those nearby bacteria that do not harbor plasmids. The bacteria with plasmids are unharmed because each plasmid also manufactures an immunity protein that protects it against the toxin of other plasmids. Therefore by killing off competing bacteria the suicidal toxin-producing gene contributes to the survival of those bacteria that carry its genetic replicas. This interpretation is supported by the fact that the plasmids tend to produce toxin when bacteria are crowded and in competition.

The most unexpected demonstration of the strength of Hamilton's theory was his use of it to explain the evolution of the social insects. Such insects live in advanced social orders characterized by cooperation, caste specialization and individual altruism. Fully social insects (insects that have all three social characteristics) exhibit a reproductive division of labor, with more or less sterile individuals (workers) laboring on behalf of fecund individuals (queens). With the single exception of the termites, all fully social insect species belong to the order Hymenoptera. The order also includes many nonsocial species, and the surprising fact is that sociality has originated on a number of separate occasions among the bees, the ants and the wasps. Hamilton was able to trace the predisposition for sociality to a particular feature of the genetic system of this order of insects.

In the hymenoptera females develop from fertilized eggs and so are diploid: they have two sets of chromosomes. Males develop from unfertilized eggs and so are haploid: they have a single set of chromosomes. In a population where both sexes are diploid, such as the one described above, the coefficient of relationship, or the average fraction of shared genes, between a mother and a daughter is the same as the coefficient of relationship between any two full siblings: in both instances r equals 1/2. As a result of the haplo-diploidy of the hymenoptera, however, a female has more genes in common with her full sisters than she has with her own daughters. Each female receives half of her genes from her haploid father and half from her diploid mother. Sisters share all of the genes they receive from their father (because he has only one set) and on the average half of the genes they receive from their mother (because she has two sets). Hence the coefficient of relationship between a mother and a daughter is still 1/2 but the coefficient of relationship between full sisters is $(1/2) \times (1) + (1/2) \times (1/2)$, or 3/4 [*see illustration on next page*].

To grasp the import of these numbers consider a species in which a female constructs and provisions a nest cell for each of her eggs and continues to lay eggs past the time when her first daughter reaches maturity. All fully social insects display an overlap of this kind. The coefficients of relationship indicate that, other things being equal, the daughter will do the most to perpetuate her genes if instead of leaving and provisioning nest cells containing her own daughters, she stays with her mother and provisions cells containing her sisters. In this way the genetic makeup of the hymenoptera predisposes them to evolve a social system in which sterile female workers care for their full siblings.

There is another twist to this argument that is still being discussed. For a female hymenopteran the coefficient of relationship is 1/2 with a son and 1/4 with a brother. Hence although

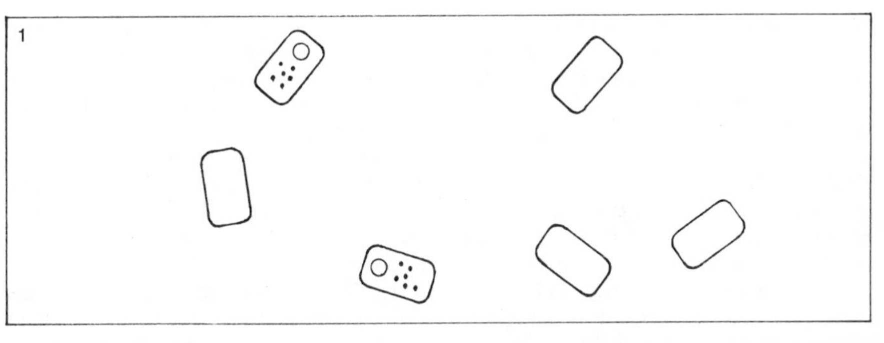

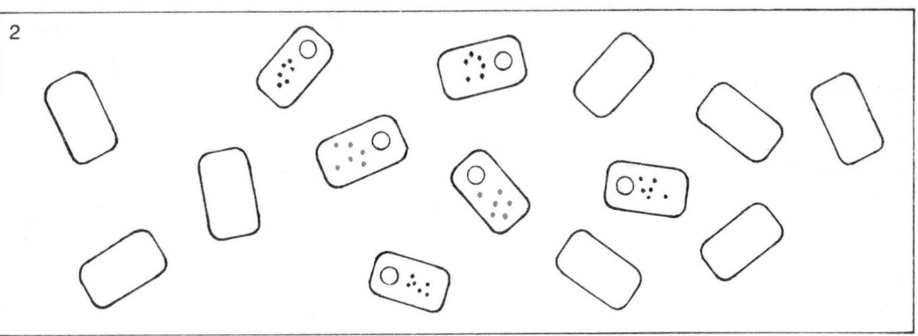

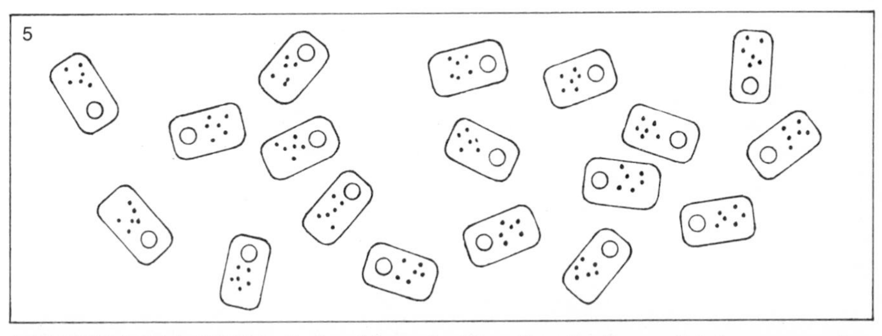

ALTRUISM OF A KIND is displayed by a viruslike self-replicating particle known as a plasmid (*open circles*), which lives as a parasite in a bacterium (*1*). Each plasmid manufactures an immunity protein (*black dots*). When the bacteria become overcrowded, some of the plasmids manufacture a colicin, or toxin (*color dots*), that kills their host bacteria and probably the plasmids themselves (*2*). When host bacteria die, colicin is released (*3*), thereby killing all nearby bacteria that do not contain the immunity protein and leaving only bacteria that are host to plasmids (*4*). The plasmids that produced colicin are destroyed, but their genetic replicas are able to multiply without competition (*5*). As example demonstrates, selection on the basis of genetic relationship operates even when individuals cannot recognize degrees of relationship.

she should raise sisters in preference to daughters, theoretically she should raise sons in preference to brothers. Hamilton points out that in fact in many social species the workers lay unfertilized eggs and raise these sons in preference to their brothers.

The matter has been taken a step further by Robert L. Trivers and H. Hare of Harvard University. If a female hymenopteran cannot distinguish between male and female eggs, she is obliged to allocate her time equally to males and females. She will therefore gain as much by raising her own offspring (r equals 1/2 for sons and r equals 1/2 for daughters) as by raising her full siblings (r equals 1/4 for brothers and r equals 3/4 for sisters). Trivers and Hare point out that in cases where the workers can distinguish the sex of the larvae they are raising they should raise an excess of females (r equals 3/4 for sisters) over males (r equals 1/4 for brothers). In fact, they showed that if the sex ratio among the reproducing members of the colony is determined by the genes in the workers, it will be about three females for every male, whereas if it is determined by genes in the queen, it will be about one female for every male. Analysis of data on ants indeed shows an excess investment in the biomass of females at the rate of about three to one. Therefore it can be deduced that the sex ratio in ants is controlled by the workers rather than by the queen. With this distorted sex ratio the worker ants should care for their siblings, male and female, in preference to their own offspring.

In recent years Hamilton's ideas have been increasingly applied to the study of the social life of higher animals, in particular birds and mammals. One such application involves the many primate species that live in groups consisting of several adult males, several adult females and their young. It is becoming apparent that in these species the young of one sex, usually the males, leave their natal troop when they reach sexual maturity and join another troop to breed. Craig Packer of the University of Sussex encountered this behavior pattern in his study of three troops of olive baboons (*Papio anubis*) in the Gombe National Reserve in Tanzania. Of 41 intertroop transfers observed over a six-year period 39 involved males, and all the males that reached maturity during that time left their natal troop.

It appears that young males leave their troop because the females in it refuse to mate with them and because the young males are attracted to unfamiliar females. This behavior seems to be selectively advantageous because a male and female born in the same troop are often close relatives and so would produce inbred offspring of low fitness. For the purposes of the present discussion,

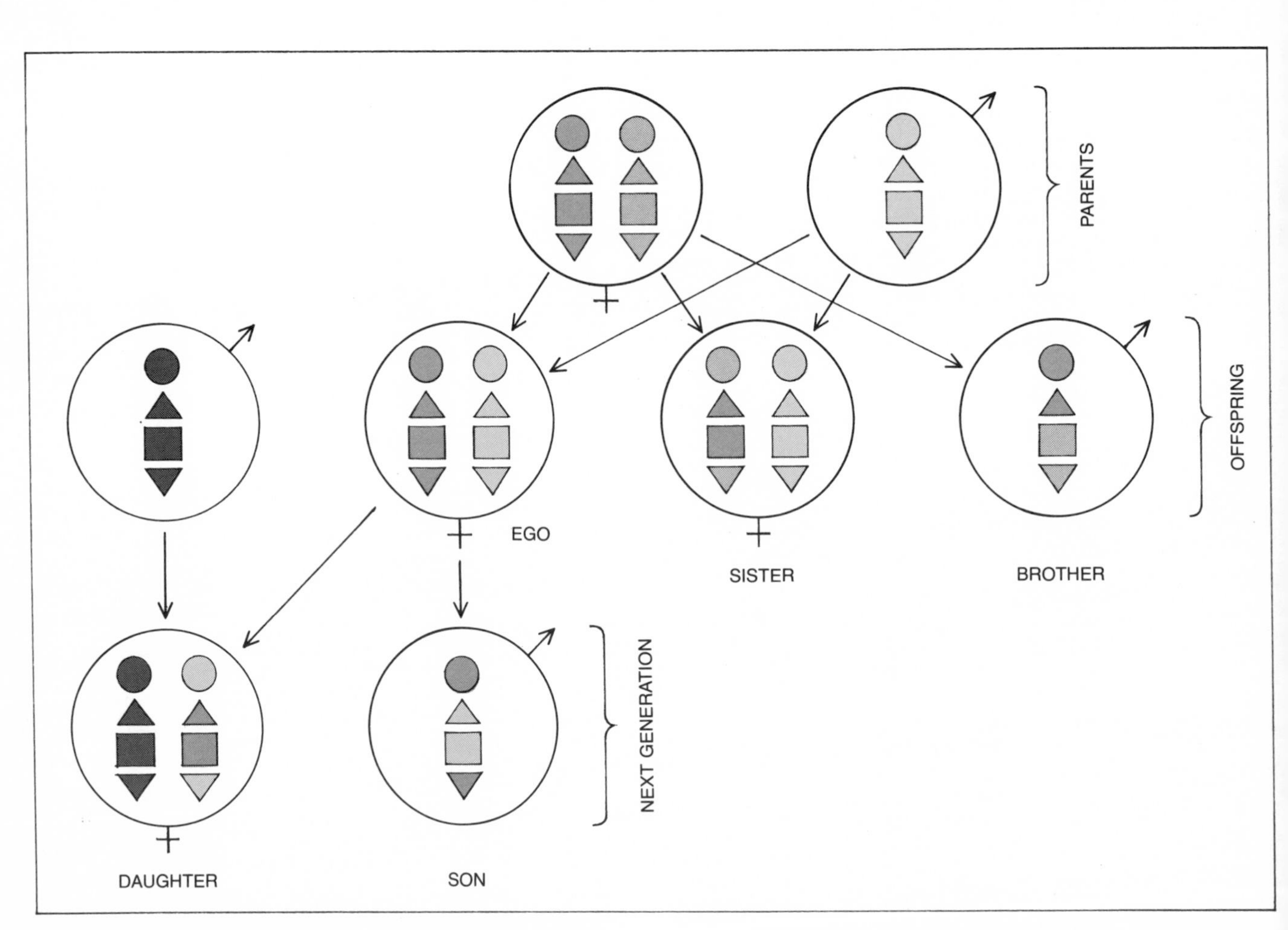

FULLY SOCIAL INSECT SPECIES are characterized by cooperation, caste specialization and individual altruism. With the exception of the termites, they all belong to the order Hymenoptera. Hamilton analyzed the frequent evolution of social behavior among these insects by examining the genetic structure of the order: females develop from fertilized eggs and have two sets of chromosomes but males develop from unfertilized eggs and have only one set of chromosomes. Consider the female Ego shown in this illustration. Any female inherits two sets of (possibly reassorted) genes: one from her mother, with two sets, and one from her father, with one set. Hence the coefficient of relationship (average fraction of shared genes) between Ego and a full sister Sib is $(1/2) \times (1/2) + (1/2) \times (1)$, or 3/4, but the coefficient of relationship between Ego and a daughter is 1/2. Ego has more genes in common with her sister than with her daughter. If Ego's mother continues to provision cells for eggs after Ego reaches maturity (all social insects display a reproductive overlap of this type), then Ego will do the most to perpetuate her genes if she helps to provision cells containing her sisters rather than provisioning cells containing her daughters. Genetic makeup of hymenoptera predisposes them to evolve social system in which sterile female workers care for siblings.

however, the most interesting aspect of this behavior pattern is not its genetic causes but its genetic effects, that is, its consequences for the genetic relationships within the troop.

As a result of the transfers the females in an olive-baboon troop will be closely related but the adult breeding males will in general not be related. In chimpanzee troops, where the males form the permanent basis of the troop and the females transfer, the situation is reversed. According to Hamilton's thesis, strong cooperation can be expected among female baboons and among male chimpanzees but not among male baboons or among female chimpanzees. The validity of this prediction is still a matter of controversy. My own guess is that it will eventually be upheld.

In spite of male baboon's lack of genetic relationship, they do display one type of cooperative behavior. When two baboons are in some kind of contest, one of them may enlist the aid of a third baboon. The soliciting baboon asks for help with an easily recognized signal, turning its head repeatedly back and forth between its opponent and its potential assistant. Packer recorded 140 instances of this type of behavior. Twenty of them involved a male *A* soliciting the help of another male *B* to take over an estrous female that was consorting with a third male *C*. In six of these instances the attempt was successful, and each time it was the soliciting male *A* that obtained the female. The fitness of the assisting male *B* does not appear to be increased by this behavior, and so the obvious question is: What does *B* get in return for its services?

The most convincing explanation of the evolution of this type of altruistic behavior between unrelated individuals is found in the concept of reciprocal altruism formulated by Trivers. According to Trivers' hypothesis, male *B*, by helping male *A* (without incurring great personal risk), gains the assurance that on a future occasion *A* will help it in return. Hence male *B* is most probably increasing its own Darwinian fitness, and the gene giving rise to this type of altruistic behavior will probably increase in frequency.

One problem with this explanation is that there appears to be no defense against cheating. What prevents male *A* from accepting help but later refusing to reciprocate? The answer may be that baboon behavior patterns have evolved so that the animals help only those individuals that do reciprocate. In that case cheating would not pay. Of course, this hypothesis presupposes individuals can recognize other individuals and remember their past behavior, but it is quite reasonable to assume that baboons possess such capabilities. Packer's data certainly support this assumption. He found that the male baboons responding most frequently to solicitations for aid also received aid most frequently and that males tended to solicit the aid of particular partners that in turn solicited aid from them.

Suppose that a population contains a small fraction p of behavioral "mutants" adopting strategy J and that the remainder of the population q adopts strategy I. If the total Darwinian fitness (expected number of surviving offspring) of the members of the population before a series of contests is C, then after the contests

$$W(I) = C + qE(I,I) + pE(I,J) \quad \text{and}$$
$$W(J) = C + qE(J,I) + pE(J,J),$$

where $E(I,J)$ is the expected payoff (change in fitness) to an individual employing strategy I in a contest with an individual employing strategy J, $W(I)$ is the total increased fitness acquired by employing strategy I and so on.

If I is an evolutionarily stable strategy, then $W(I) > W(J)$ for any mutant strategy J. In this case

$$\text{either } E(I,I) > E(J,I) \quad \text{or}$$
$$E(I,I) = E(J,I) \quad \text{and} \quad E(I,J) > E(J,J).$$

GAME-THEORY MODELS help to explain the use of conventions in animal contests, another type of behavior that does not seem to promote the survival of the individual displaying it. For each model there is sought an evolutionarily stable strategy, that is, a strategy that confers the highest reproductive fitness on the animals adopting it. This illustration shows the mathematical requirements for a strategy to be evolutionarily stable. More generally, an evolutionarily stable strategy can be defined as a strategy with the property that if all members of a population adopt it, then no mutant strategy can invade the population. It is assumed that members of model population engage in contests in random pairs and that subsequently each individual reproduces in proportion to payoff (change in Darwinian fitness) it has accumulated. It now appears that many types of conventional fighting are indeed evolutionarily stable strategies.

Over the past few years I have become particularly interested in the evolution of a type of ritualized animal behavior: the use of conventions in animal contests. Animals engaged in a contest over some valuable resource (such as a mate, territory or position in a hierarchy) do not always use the weapons available to them in the most effective way. They may instead act according to certain conventions (employing threat displays, refraining from attacking an opponent in a vulnerable position and so on), often pursuing a kind of limited warfare that avoids serious injury. For example, when male fiddler crabs fight over the possession of a burrow, they use a powerful enlarged claw as a weapon. Although the claw is strong enough to crush the abdomen of an opponent, no crab has ever been known to injure another in such a fight. (It would be wrong to deduce from this example that animals are never injured in intraspecific fights or that animals never fight to the death. The conventional behavior is sufficiently common, however, to require an explanation.)

At the time when I first learned of the problem the evolution of conventional fighting was explained by arguing that if intraspecific fighting were not conventional, then a great many animals would be injured. In other words, conventional behavior evolved because unconventional behavior would, as Julian Huxley had put it, "militate against the survival of the species." As a student of Haldane's I had been taught to be distrustful of arguments that depend on "the good of the species." This particular one did not seem to be able to account for the complex anatomical and behavioral adaptations for limited conflict found in many species. I thought there should be a way to explain how natural selection operates on the individual to promote those characteristics, that is, to show that conventional behavior increases the Darwinian fitness of the individual displaying it.

It appeared, however, that an individual's fitness would be increased not by conventional fighting but by unconventional fighting. It seemed to me that, in a contest between individuals *A* and *B*, if *A* obeyed the rules and *B* "hit below the belt," then *B* would win the contest and pass its genes on to the next generation. This puzzle remained at the back of my mind until 1970, when an unpublished paper by G. R. Price prompted me to review it. It occurred to me then that I might gain some understanding of the problem by borrowing some of the concepts of the mathematical theory of games.

Game theory was formulated by John von Neumann and Oskar Morgenstern in the 1940's for the purpose of analyzing human conflict. The theory in particular seeks to determine the optimum strategy to pursue in conflict situations. I hoped that by applying a modified form of game theory I would be able to construct a mathematical model of animal contests and so determine what strategies would be favored by natural selection at the level of the individual. If all went well, experimental evidence and observation would support the mathematically derived conclusions.

The strategies I was seeking had little

to do with the optimum strategies with which traditional game theory deals. For each game model of animal contests I hoped to determine an evolutionarily stable strategy: a strategy with the property that if most of the members of a large population adopt it, then no mutant strategy can invade the population. In other words, a strategy is evolutionarily stable if there is no mutant strategy that gives higher Darwinian fitness to the individuals adopting it.

Consider a simple model: a species that in contests between two individuals has only two possible tactics, a "hawk" tactic and a "dove" one. A hawk fights without regard to any convention and escalates the fighting until it either wins (that is, until its opponent runs away or is seriously injured) or is seriously injured. A dove never escalates; it fights conventionally, and then if its opponent escalates, it runs away before it is injured.

At the end of a contest each contestant receives a payoff. The expected payoff to individual X in a contest with individual Y is written $E(X,Y)$. The payoff is a measure of the change in the fitness of X as a result of the contest, and so it is determined by three factors: the advantage of winning, the disadvantage of being seriously injured and the disadvantage of wasting time and energy in a long contest. For the hawk-dove game suppose the effect on individual fitness is $+10$ for winning a contest and -20 for suffering serious injury. Suppose further two doves can eventually settle a contest but only after a long time and at a cost of -3. (The exact values of the payoffs do not affect the results of the model as long as the absolute, or unsigned numerical, value of injury is greater than that of victory.)

The game can be analyzed as follows. If the two individuals in a contest both adopt dove tactics, then since doves do not escalate, there is no possibility of injury and the contest will be a long one. Each contestant has an equal chance of winning, and so the expected payoff to one of the doves D equals the probability of D winning the contest (p equals $1/2$) times the value of victory plus the cost of a long battle, that is, $E(D,D)$ equals $(1/2)(+10) + (-3)$, or $+2$. Similarly, a hawk fighting another hawk has equal chances of winning or of being injured but in any case the contest will be settled fairly quickly. Hence the expected payoff $E(H,H)$ is equal to $(1/2)(+10) + (1/2)(-20)$, or -5. A dove fighting a hawk will flee when the hawk escalates, so that the dove's expected payoff is 0 and the victorious hawk's payoff is $+10$.

Now suppose the members of a population engage in contests in the hawk-dove game in random pairs and subsequently each individual reproduces its kind (individuals employing the same strategy) in proportion to the payoff it has accumulated. If there is an evolutionarily stable strategy for the game, the population will evolve toward it. The question, then, is: Is there an evolutionarily stable strategy for the hawk-dove game?

It is evident that consistently playing hawk is not an evolutionarily stable strategy: a population of hawks would not be safe against all mutant strategies. Remember that in a hawk population the expected payoff per contest to a hawk $E(H,H)$ is -5 but the expected payoff to a dove mutant $E(D,H)$ is 0. Hence dove mutants would reproduce more often than hawks. A similar argument shows that consistently playing dove is also not an evolutionarily stable strategy.

There is a precise mathematical definition for an evolutionarily stable strategy: A strategy I is evolutionarily stable if, for any mutant strategy J, either $E(I,I)$ is greater than $E(J,I)$ or $E(I,I)$ equals $E(J,I)$ and $E(I,J)$ is greater than $E(J,J)$. Although neither of the pure strategies labeled "Always play hawk" or "Always play dove" fulfills one of these requirements, there is a mixed strategy that does. A mixed strategy is one that prescribes different tactics to be followed in a game according to a specified probability distribution. The mixed strategy that is evolutionarily stable for the hawk-dove game is play hawk with probability $8/13$ and play dove with probability $5/13$. I shall not discuss the derivation of this strategy here, but it is not difficult to see that the strategy does fulfill the second requirement of being evolutionarily stable against, say, a mutant hawk strategy.

If the mixed strategy is called M, it will suffice to show that $E(M,M)$ equals $E(H,M)$ and $E(M,H)$ is greater than $E(H,H)$. This can be done by applying the definition of strategy M: the payoff $E(M,M)$ is equal to $(8/13)E(H,M) + (5/13)E(D,M)$, and $E(H,M)$ is equal to $(8/13)E(H,H) + (5/13)E(H,D)$ and $E(D,M)$ is equal to $(8/13)E(D,H) + (5/13)E(D,D)$. The values already computed for the hawk-dove game can now be substituted into these equations so that $E(M,M)$ is equal to $(8/13)(10/13) + (5/13)(10/13)$, or $10/13$. The preceding calculation showed that $E(H,M)$ equals $10/13$, and so $E(M,M)$ and $E(H,M)$ are equal. Furthermore, the payoff $E(M,H)$ equals $(8/13)E(H,H) + (5/13)E(D,H)$, or $-40/13$, and $E(H,H)$ equals -5, and so $E(M,H)$ is greater than $E(H,H)$. In other words, the strategy hawk cannot invade a population employing the mixed strategy M.

The hawk-dove model predicts that mixed strategies will be found in real animal contests, either in the form of different animals adopting different tactics (such as hawk and dove) or in the form of individuals varying their tactics. Animal behavior in many contest situations is indeed variable, but of course that does not prove that an evolutionari-

SERIOUS INJURY = −20
VICTORY = +10
LONG CONTEST = −3

$E(H,H) = ½(+10) + ½(-20) = -5$
$E(H,D) = +10$
$E(D,H) = 0$
$E(D,D) = ½(+10) + (-3) = +2$

	HAWK (H)	DOVE (D)
HAWK (H)	−5	+10
DOVE (D)	0	+2

IN THE HAWK-DOVE GAME, illustrated here, there are only two tactics that can be employed in contests between two individuals: a "hawk" tactic and a "dove" one. A hawk fights without regard to any convention and escalates a contest until it wins or is seriously injured. A dove fights conventionally, never escalating; if its opponent escalates, it runs away before it is injured; two doves can settle a contest but only after a long period of time. The changes in a contestant's Darwinian fitness as a result of serious injury, of a long contest and of victory are shown at the upper left in the illustration. (The exact values of these factors do not affect the model results as long as the unsigned numerical value of injury is greater than that of victory.) Calculations of the expected payoffs to individuals in different contests are shown at the upper right. The payoffs are displayed in the matrix at the bottom. Each payoff is to the individual employing the tactic directly to the left in the matrix in a contest with an individual employing the tactic directly above. For example, +10 (*color*) equals $E(H,D)$, the expected payoff to a hawk H in contest with a dove D. In hawk-dove game neither of pure strategies designated "Always play hawk" or "Always play dove" is evolutionarily stable. Only evolutionarily stable strategy is mixed strategy: play hawk with probability 8/13 and play dove with probability 5/13.

ly mixed strategy is operating. One case of animal behavior that does conform rather well to the model is found in investigations into the behavior of the dung fly conducted by G. A. Parker of the University of Liverpool.

Female dung flies lay their eggs in cowpats, and so males congregate at cowpats and try to mate with the arriving females. Parker found that the rate at which the females arrive at a cowpat decreases as the cowpat gets stale. In game terms the male is presented with a choice of two tactics as the cowpat he is patrolling gets stale. He can leave in search of a fresh cowpat or he can stay. The success of the male's choice of tactic of course depends on the behavior of other males. If most of the other males leave as soon as the cowpat gets stale, then he should stay, because although relatively few females will be arriving, he will have little or no competition in mating with them. On the other hand, if the other males stay, then he should leave. In other words, the only evolutionarily stable strategy is a mixed one in which some males leave early and others stay. Game-theory analysis predicts that with this strategy when the system reaches an equilibrium, early-leaving and late-leaving males should have the same average mating success. Parker's data yield precisely that result. It is not known, however, whether the evolutionarily stable mixed strategy of the dung fly is achieved by some males' consistently leaving early and others' consistently leaving late or by individual males' varying their tactics.

SERIOUS INJURY = −20
VICTORY = +10
LONG CONTEST = −3

$$E(H,B) = \tfrac{1}{2}E(H,H) + \tfrac{1}{2}E(H,D) = -\frac{5}{2} + \frac{10}{2} = +2.5$$

$$E(D,B) = \tfrac{1}{2}E(D,H) + \tfrac{1}{2}E(D,D) = +0 + \frac{2}{2} = +1$$

$$E(B,H) = \tfrac{1}{2}E(H,H) + \tfrac{1}{2}E(D,H) = -\frac{5}{2} + 0 = -2.5$$

$$E(B,D) = \tfrac{1}{2}E(H,D) + \tfrac{1}{2}E(D,D) = +\frac{10}{2} + \frac{2}{2} = +6$$

$$E(B,B) = \tfrac{1}{2}E(H,D) + \tfrac{1}{2}E(D,H) = +\frac{10}{2} + 0 = +5$$

	HAWK (*H*)	DOVE (*D*)	BOURGEOIS (*B*)
HAWK (*H*)	−5	+10	+2.5
DOVE (*D*)	0	+2	+1
BOURGEOIS (*B*)	−2.5	+6	+5

HAWK-DOVE-BOURGEOIS GAME, illustrated here, models animal contests that are characterized by uncorrelated asymmetries, that is, differences between contestants that do not necessarily affect the outcome or payoffs of the contests. Asymmetries of this type often serve to settle real contests conventionally. A contest over some resource between the owner of the resource and an interloper is a good example of an uncorrelated asymmetry, and so it was used to define a new tactic, "bourgeois," to be added to the tactics in the hawk-dove game. If a bourgeois contestant is the owner of the resource in question, it adopts the hawk tactic; otherwise it adopts the dove tactic. It is assumed that each contest is between an owner and an interloper, that each individual is equally likely to be in either role and that each individual knows which role it is playing. Pure strategy "bourgeois" is the only evolutionarily stable strategy for the game. There can never be an escalated contest between opponents adopting strategy, since one will be owner and playing hawk and other will be interloper and playing dove. Hence ownership is taken as conventional cue for settling contests in model population. Many examples of bourgeois strategy have been found in real animal populations (*see illustration on next two pages*).

It is obvious that real animals can adopt strategies that are more complex than "Always escalate," "Always display" or some mixture of the two. For example, some animals make probes, or trial escalations. Others employ conventional tactics but will escalate in retaliation for an opponent's escalation. There is, however, another important way in which many real animal contests do not conform to the hawk-dove model. Most real contests are asymmetric in that, unlike hawks and doves, the contestants differ from each other in some area besides strategy.

Three basic types of asymmetries are encountered in animal contests. First, there are asymmetries in the fighting ability (the size, strength or weapons) of the contestants; differences of this kind are likely to affect the outcome of an escalated fight. Second, there are asymmetries in the value to the contestants of the resource being competed for (as in a contest over food between a hungry individual and a well-fed one); differences of this kind are likely to affect the payoffs of a contest. Third, there are asymmetries that are called uncorrelated because they affect neither the outcome of escalation nor the payoffs of a contest. For the purposes of this discussion the uncorrelated asymmetries are of special interest because they often serve to settle contests conventionally.

Perhaps the best example of an uncorrelated asymmetry is found in a contest over a resource between the "owner" of the resource and an interloper. In calling this an uncorrelated asymmetry I do not mean that ownership never alters the outcome of escalation or the payoffs of contests; I simply mean that ownership will serve to settle contests even when it does not alter those factors. To demonstrate the effect of such an uncorrelated asymmetry I shall return to the hawk-dove game and add to it a third strategy called bourgeois: if the individual is the owner of the resource in question, it adopts the hawk tactic; otherwise it adopts the dove tactic.

In the hawk-dove-bourgeois game it is assumed that each contest is between an owner and an interloper, that each individual is equally likely to be in either role and that each individual knows which role it is playing. The payoffs for contests involving hawks and doves are unchanged by the addition of the new strategy, but additional payoffs must be calculated for contests that involve bourgeois contestants [*see illustration above*]. For example, in a contest between a bourgeois and a hawk there is an equal chance that the bourgeois will be the owner (and so playing hawk) or the interloper (and so playing dove); hence $E(B,H)$ equals $(1/2)E(H,H) + (1/2)E(D,H)$, or $(1/2)(-5) + (1/2)(0)$, or -2.5. The remaining payoffs are calculated in a similar manner. The main point, however, is that there can never be an escalated contest between two opponents playing bourgeois, because if one is the owner and playing hawk, then the other is the interloper and playing dove. Therefore the payoff $E(B,B)$ is equal to $(1/2)E(H,D) + (1/2)E(D,H)$, or $(1/2)(10) + (1/2)(0)$, or 5. When this figure is compared with the other payoffs, it is not difficult to see that consistently playing bourgeois is the only evolutionarily stable strategy for this game. Thus ownership is taken as a conventional cue for settling contests.

Hans Kummer of the University of Zurich has observed a beautiful example of the bourgeois strategy in the hamadryas baboon (*Papio hamadryas*). In this species a single male forms a permanent bond with one or more females and is not normally challenged by other males. Kummer performed the following experiment with three unacquainted baboons. Male *A* and a female were put in an enclosure and male *B* was put in a cage from which he could see what was happening in the enclosure but could not interfere. In a relatively short time (about 20 minutes) a bond was established between male *A* and the female. Male *B* was then released into the enclosure. He did not attempt to annex the female and in fact avoided any kind of confrontation with *A*.

There are two possible explanations for male *B*'s behavior. It may be that, as the model predicts, ownership is taken as a conventional cue for the settlement of contests. On the other hand, male *B* may have perceived that male *A* was stronger and would probably have won an escalated contest. Kummer was able to eliminate the second possibility by repeating the experiment with the same two males and a different female several weeks later. Now, however, the roles of the male baboons were reversed and *B* was placed in the enclosure with the female and *A* was placed in the cage. This time it was *B* that annexed the female and was not challenged by *A*. The bourgeois principle was indeed operating. (It should be noted that there is more to the story; since these experiments were done Kummer has found that female preference also plays a role.)

N. B. Davies of the University of Oxford has discovered another example of the bourgeois strategy in the speckled wood butterfly (*Pararge aegeria*). Males of the species claim and defend sunlit spots on the forest floor, where they can court more females than they can in the forest canopy. There are never enough spots for all the males to occupy at any one time, and so there are always males patrolling the forest canopy. On occasion an interloping male flies into an occupied sunlit spot and is challenged by the owner. The two males then make a brief spiral flight up toward the canopy, after which one flies up into the canopy and the other settles back down into the spot. By marking male butterflies Davies was able to show that it is invariably the original owner that returns to the sunlit spot after a spiral flight.

Once again there are two plausible explanations for such behavior. It is possible that ownership is accepted as a cue and that the spiral flight serves somehow to inform the interloper that the sunlit spot is occupied. Or it is possible that only relatively strong butterflies hold sunlit spots and that the spiral flight serves to demonstrate their strength. Davies favors the first explanation: that the speckled wood butterfly is operating according to the bourgeois principle. He has two reasons for doubting that only strong butterflies hold sunlit spots. To begin with, he noted that most of the males he marked in the canopy were eventually observed holding spots. Moreover, he performed an experiment in which he removed the owner from a

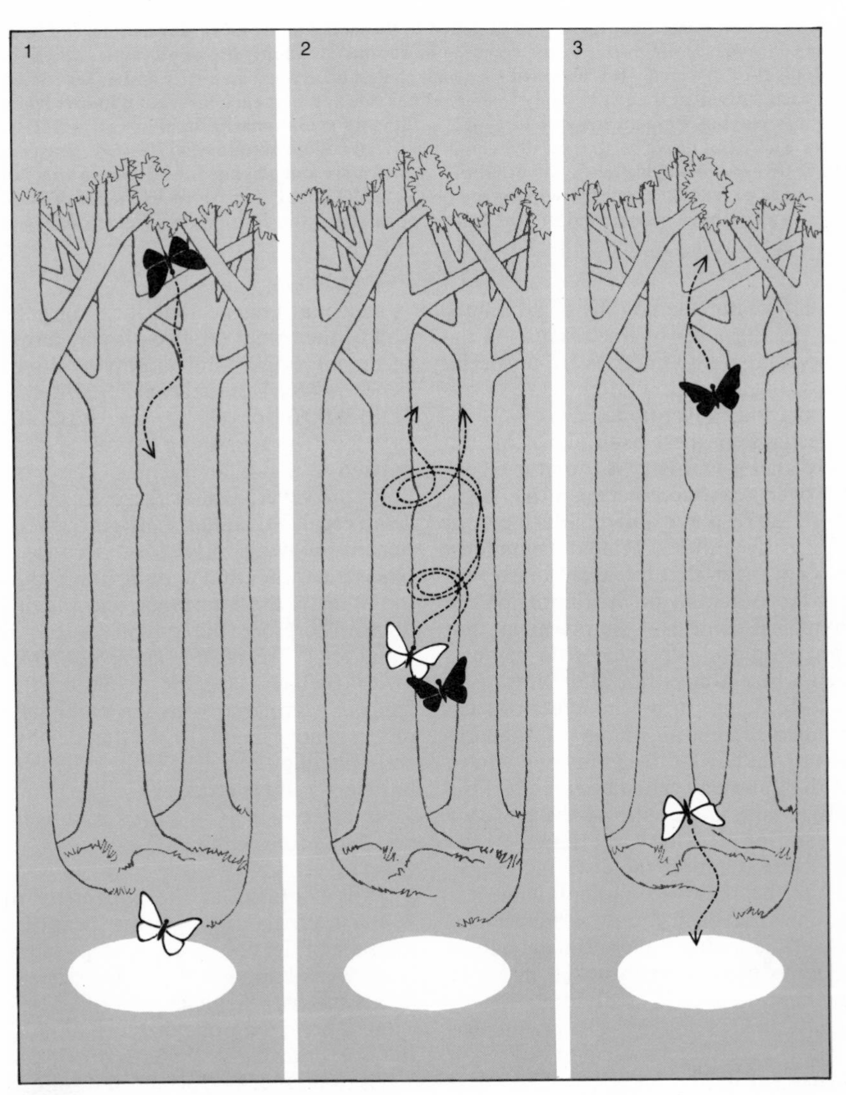

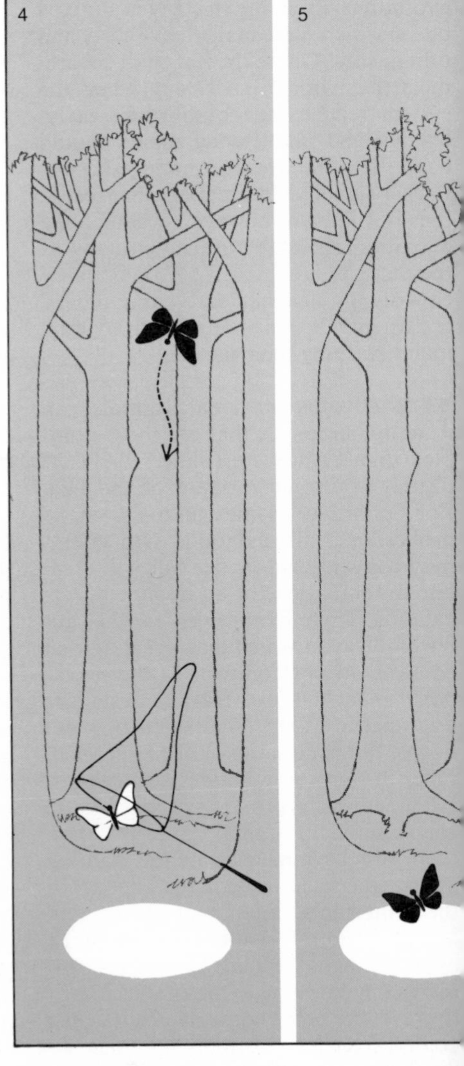

EXAMPLE OF BOURGEOIS STRATEGY has been discovered by N. B. Davies of the University of Oxford. Males of the speckled wood butterfly (*Pararge aegeria*) follow the bourgeois strategy in territorial disputes over sunlit spots on the forest floor. If a male from the forest canopy descends into an occupied sunlit spot (*1*), it is challenged by the owner of the spot. The two butterflies execute a short spiral fligh up toward the forest canopy (*2*), after which the original owner return to the sunlit spot and the interloper returns to the canopy (*3*). Ther is additional evidence to suggest that ownership is accepted as a co ventional cue in this species and that the spiral flight serves someho

sunlit spot, waited until a new male descended from the canopy and then released the original owner back into the spot. On each occasion the new owner won the ensuing dispute and the original owner retreated.

This last experiment suggests that a male of this species considers that it owns a sunlit spot when it has settled unchallenged in the spot for a few seconds. What happens if two males consider themselves to be owners of the same spot? Davies investigated the question by surreptitiously introducing a second male into occupied territory. Sooner or later one of the owners would notice the other and challenge it. In all cases there ensued a protracted spiral flight lasting an average of 10 times longer than a normal flight. It appears that a butterfly perceiving itself to be an owner is prepared to escalate.

Not all asymmetric contests are as simple as the ones I have described so far. For example, contests between male fiddler crabs of the species *Uca pugilator* can involve an asymmetry between the owner of a burrow and a wandering crab and also asymmetries in the crabs' size and strength. It appears that a large component of the behavior of the crabs is concerned with assessing the asymmetries. Gary W. Hyatt and Michael Salmon of the University of Illinois found that in 403 fights between males of this species the burrow owner prevailed in 349 instances. In the 54 fights won by the wandering male that crab was larger than the burrow owner in 50 instances and smaller in only one. It is evident that asymmetries in ownership and fighting ability are relevant, but it will not be an easy task to develop and test a game-theory model that can help to explain the form and duration of the contests.

There are many relevant factors in animal contests that I have not discussed. For example, in some instances the contested resource is divisible, so that sharing it may be preferable to fighting over it. Some animals provide false information about their size (as with a ruff or a mane) or their intentions. Game-theory models will have to be devised that take account of these features. There are also different types of animal behavior for which game-theory models are appropriate. For example, parental care is not normally regarded as a contest because parents have a common interest in the survival of their offspring. The activity has areas of conflict as well as areas of common interest, however, and I believe a game-theory analysis is illuminating. Finally, in seeking the differences and similarities between man and other animals, it may be helpful to analyze the "games" they can play.

to inform the interloper that the sunlit spot is occupied. If the owner is removed from a spot (*4*), another male will descend to occupy it (*5*). When the original owner is placed back in the spot (*6*), a spiral flight ensues (*7*), and this time it is the new owner that returns to occupy the spot (*8*). If two male butterflies consider themselves to be owners of a single sunlit spot, one challenges the other and they execute an extremely long spiral flight (*9*) from which either one can emerge the winner (*10*). It appears that, just as the hawk-dove-bourgeois game predicts, a male speckled wood butterfly that considers itself to be the owner of a sunlit spot is willing to escalate a contest.

VIII

The Evolution of Man

The Evolution of Man

BY SHERWOOD L. WASHBURN

A wealth of new fossil evidence indicates that manlike creatures had already branched off from the other primates by four million years ago. Homo sapiens himself arose only some 100,000 years ago

Perhaps the most significant single fact about human evolution has a paradoxical quality: the brain with which man now begins to understand his own lengthy biological past developed under conditions that have long ceased to exist. That brain evolved both in size and in neurological complexity over some millions of years, during most of which time our ancestors lived under a daily obligation to act and react on the basis of exceedingly limited information. What is more, much of the information was wrong.

Consider what that meant. Before the information being fed to man's evolving brain began to be refined by an advancing technology our ancestors lived in a world that seemed to them small and flat and that they could assess only in very personal terms. Sharing the world with them were divine spirits, ghosts and monsters. Yet the brain that developed these concepts was the same brain that today deals with the subtleties of modern mathematics and physics. And it is this same technological progress that allows us to recognize human evolution.

One relevant example of the paradox is the extraordinary expansion over the past two centuries of man's perception of time. As Ernst Mayr points out in the introductory article of this *Scientific American* book, at the beginning of the 18th century the accepted view was that the time that had elapsed between the creation of the earth and the present was no more than a few thousand years. At the end of the 19th century the perceived interval had been enlarged a thousandfold and stood at about 40 million years. With the discovery that the slow and constant decay of certain radioactive isotopes constitutes a clock it became necessary to enlarge the interval another hundredfold, so that today we reckon the age of the earth to be about 4.6 billion years.

The human mind cannot literally comprehend such an interval; it is as ineffable as the trillions and quadrillions of dollars that are juggled in world economics. Man's common sense perceives time as being short: a rhythm of birth, growth and death. To this sense of biological time can be added a sense of social time: a less tangible interval of three to five generations that is important to the actors in the drama of human society. Longer intervals do not have the same emotional impact. The real time scale of the universe that has been developed by science can be regarded as having liberated the perception of time from the limitations of the human mind.

The modern perception of time is of course only one in a series of mental emancipations that have led man to a deeper understanding of his own evolutionary history. In what follows I shall review information bearing on human evolution that has been gathered by workers in several fields of study, and I shall also assess the contribution of each field to our overall grasp of the subject today. No one can be an expert in all these fields, and this article might better be regarded as a personal evaluation than as an objective summary.

Few things defy common sense more than the concept that the continents of the earth are constantly in motion and that their positions have shifted drastically on a time scale short compared with the age of the earth. One consequence of yesterday's common sense is that all traditional theories of human evolution have assumed that the positions of the continents are fixed. To be sure, "land bridges" between continents and shallow seas that invaded continents were postulated, but the positions of the great continental plates were not affected. Although it has been nearly 70 years since Alfred Wegener proposed that continental drift was a reality, it is only over the past 20 years, as the mechanism of plate tectonics has been developed and the movement of continents has actually been measured, that the concept of moving continents has become accepted and even respectable.

The combined data from radioactive-isotope dating and plate tectonics have fundamentally changed the background of human evolutionary studies. For example, everyone used to think that the monkeys of the New World had evolved directly from the primitive prosimians that had once flourished in North America. (The prosimians are the least advanced of the order Primates, which includes the apes and man. All living prosimians are confined to the Old World.) We now know, however, that 35 to 40 million years ago Africa was as close to South America as North America was. Some of the primates that were ancestral to the New World monkeys might just as easily have been accidentally rafted (perhaps on a tree felled by a flood) to South America from Africa as from North America. The propinquity of the three continents, which was unthinkable before continental drift was accepted, does not prove that the ancestral stock of the New World monkeys emigrated from Africa; nevertheless, it does present that entirely new and important possibility.

Another example of the effect of plate tectonics on human evolutionary hypotheses has to do with recent and repeated assertions that Africa was the focus of human origins. What the history of continental drift indicates is that there were broad connections between Africa and Eurasia from the time of their collision some 18 million years ago (when some ancestral elephants left Africa and spread over Eurasia) until the flooding of the Mediterranean basin five or six million years ago. It happens that

IMMUNOLOGICAL DISTANCES between selected mammals are indicated by the separation, as measured along the horizontal axis, between the branches of this "divergence tree." For example, the monotremes (primitive egg-laying mammals) are removed from the marsupials by a distance (in arbitrary units) of only 1.5 but are removed from the chimpanzee by a distance of nearly 17. The distance between man and the Old World monkeys is a little more than 3, between man and the Asiatic gibbons 2, and between man and the gorillas and the chimpanzees of Africa less than 1. The data are from Morris Goodman of Wayne State University.

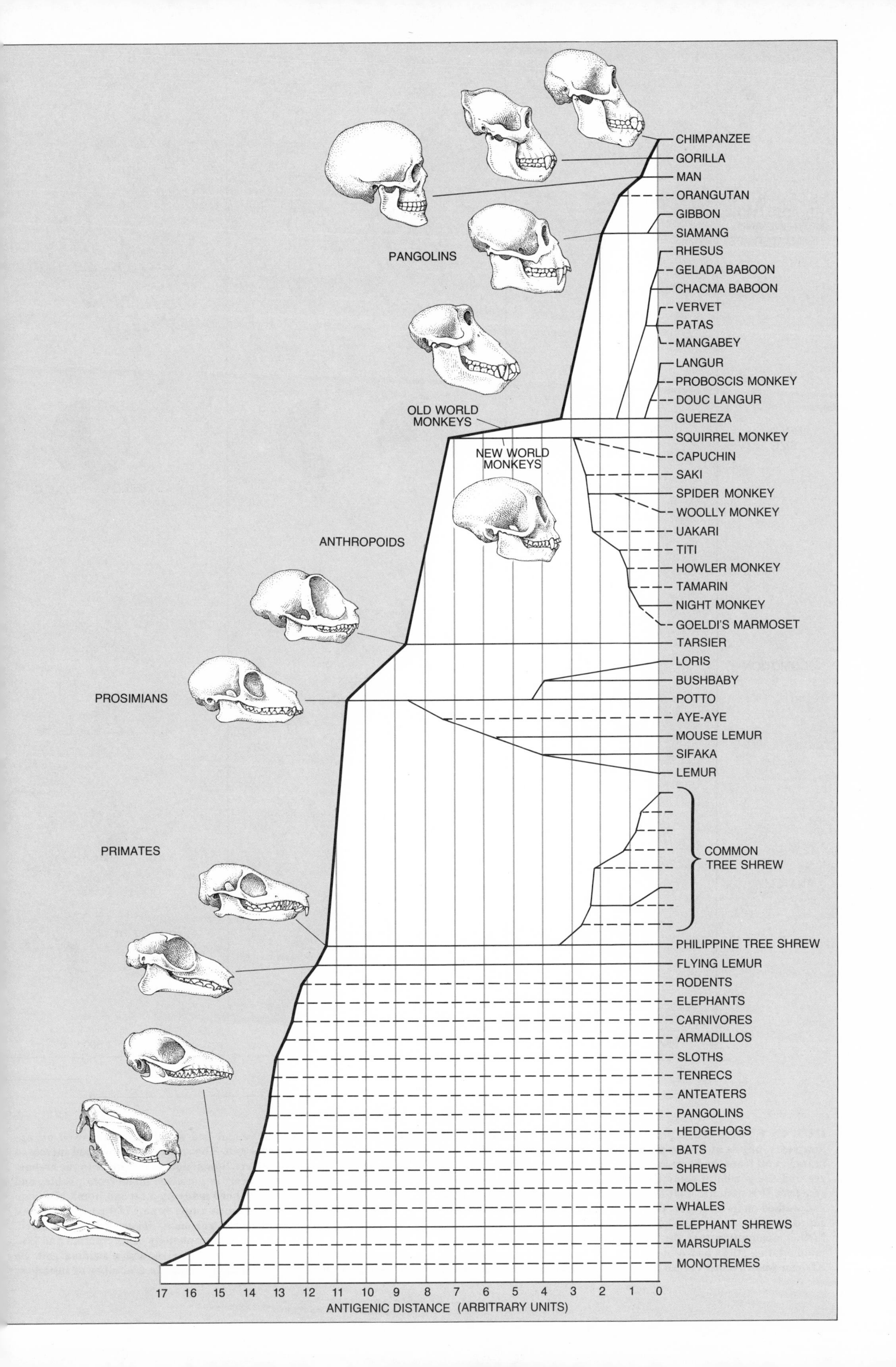

CHIMPANZEE
GORILLA
MAN
ORANGUTAN
GIBBON
SIAMANG
RHESUS
GELADA BABOON
CHACMA BABOON
VERVET
PATAS
MANGABEY
LANGUR
PROBOSCIS MONKEY
DOUC LANGUR
GUEREZA
SQUIRREL MONKEY
CAPUCHIN
SAKI
SPIDER MONKEY
WOOLLY MONKEY
UAKARI
TITI
HOWLER MONKEY
TAMARIN
NIGHT MONKEY
GOELDI'S MARMOSET
TARSIER
LORIS
BUSHBABY
POTTO
AYE-AYE
MOUSE LEMUR
SIFAKA
LEMUR
COMMON TREE SHREW
PHILIPPINE TREE SHREW
FLYING LEMUR
RODENTS
ELEPHANTS
CARNIVORES
ARMADILLOS
SLOTHS
TENRECS
ANTEATERS
PANGOLINS
HEDGEHOGS
BATS
SHREWS
MOLES
WHALES
ELEPHANT SHREWS
MARSUPIALS
MONOTREMES
PANGOLINS
OLD WORLD MONKEYS
NEW WORLD MONKEYS
ANTHROPOIDS
PROSIMIANS
PRIMATES
17 16 15 14 13 12 11 10 9 8 7 6 5 4 3 2 1 0
ANTIGENIC DISTANCE (ARBITRARY UNITS)

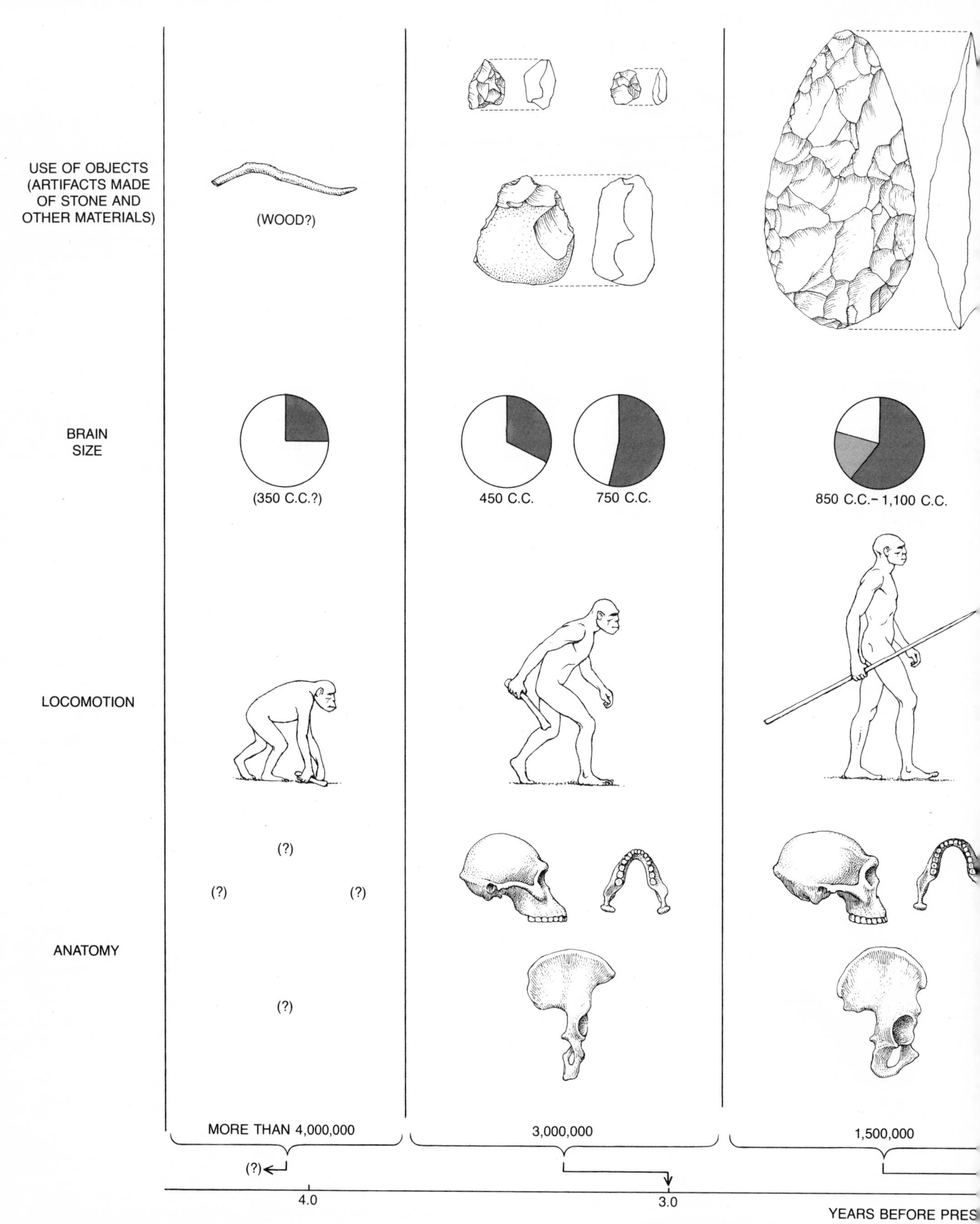

HUMAN EVOLUTION, projected over a possible span of 10 million years, begins at a slow pace when a still undiscovered hominid branches off from the hominoid stock ancestral to man, the chimpanzee and the gorilla at some time more than four million years ago (*far left*). It is assumed that the ancestral hominid had a small brain and walked on its knuckles. This mode of locomotion enables a quadruped to move about while holding objects in its hands, leading to the further assumption that the hominid outdid living chimpanzees in manipulating sticks and other objects. By four million years ago the African fossil record reveals the presence of an advanced hominid: *Australopithecus*. This subhuman had a pelvis that allowed an upright posture and a bipedal gait. The size of the brain had increased to some 450 cubic centimeters. Stone tools soon appear in the archaeological record; they are simple implements made from pebbles and cobbles. The tools may have been made by a second hominid group, chiefly notable for having a much larger brain: 750 c.c. Next, about 1.5 million years ago, the first true man, *Homo erectus,* appeared. Still primitive with respect to the morphology of its cranium and jaw, *H. erectus* had an essentially modern pelvis and a striding gait. Its brain size approaches the modern average in a number of instances.

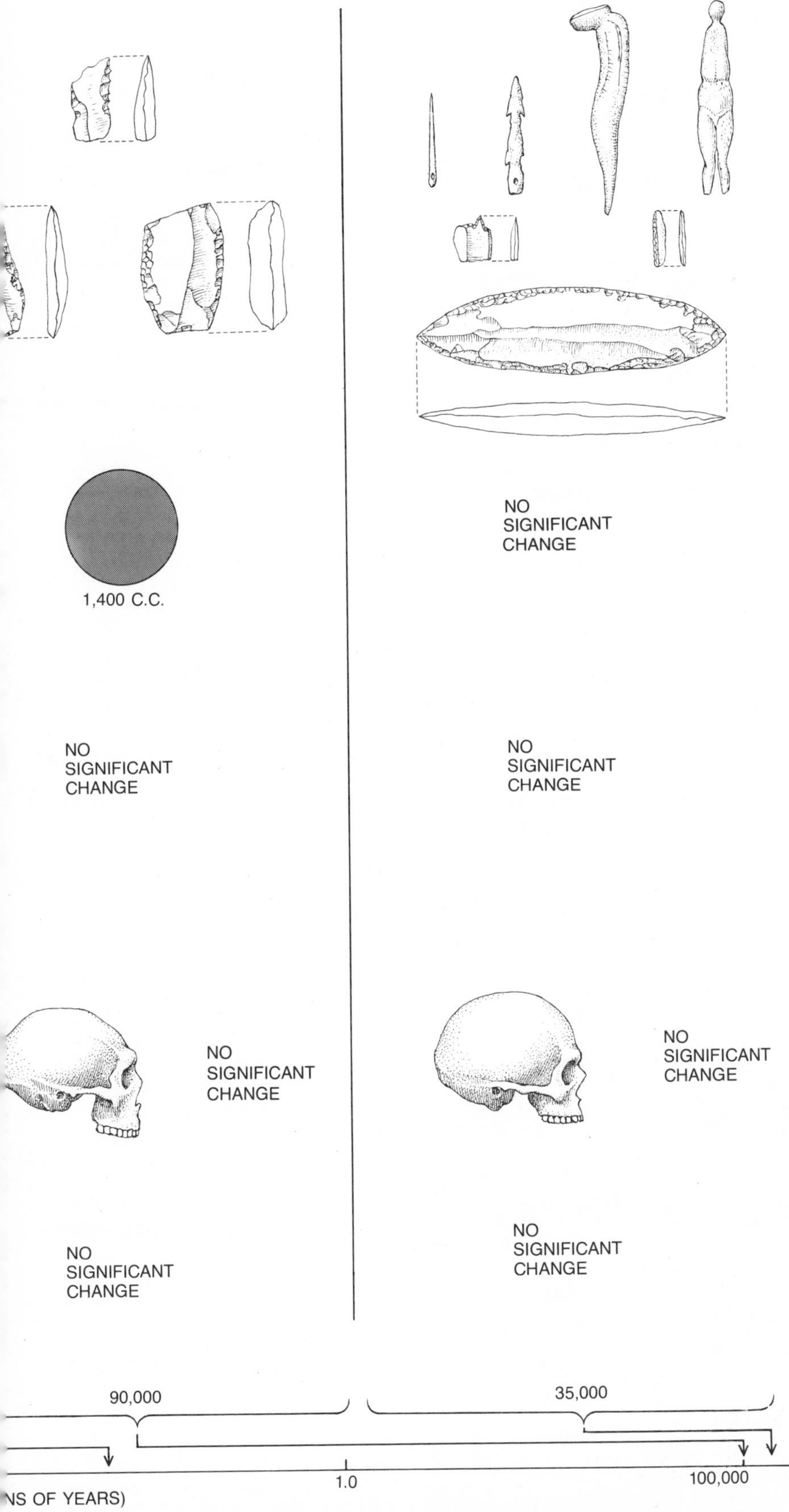

Many stone tools that are contemporaneous with the fossils of *H. erectus* are "cores" from which flakes have been removed on two sides; they are representative of the Acheulian tool industry. Not until some 100,000 years ago did *Homo sapiens* appear, in the form of Neanderthal man. The shape of Neanderthal's skull is not quite modern but the size of its brain is. Most of the tools found at Neanderthal sites represent the Mousterian industry; they are made from flakes of flint rather than cores. Only 40,000 years ago modern man, *Homo sapiens sapiens,* arrived on the scene. His skull is less robust than that of Neanderthal and his brain is slightly smaller. Many of his stone tools are slender blades; some, known as laurel-leaf points, appear to be ceremonial rather than utilitarian. Among his bone artifacts are needles, harpoon heads, awls and statuettes. About 10,000 years ago man's transition from hunting to farming began.

the fossil remains of *Ramapithecus,* the Miocene-Pliocene ape commonly believed to be an ancestor of the hominid line, that is, the line of man and his extinct close relatives, are found from India and Pakistan through the Near East and the Balkans to Africa. The continuity in the distribution of these fossils suggests that the geography of Eurasia and Africa then was substantially different from what it is today. Moreover, the list of identical Indian and African faunas can be extended far beyond this single extinct ape: both regions harbor macaque monkeys, lions, leopards, cheetahs, jackals, wild dogs and hyenas. The possibility that man originated solely in Africa therefore seems less likely than it once did. In other words, the longer man's ancestors existed as intelligent, upright-walking, tool-using hunters, the less likely it is that their distribution was confined to any one continent.

Comparative anatomy is a field of study considerably older than plate tectonics. Its roots are in the 19th century, and it has been the discipline most concerned with the similarities and differences between man and his fellow primates. Its basic assumption has been that a sufficiently large quantity of information will inevitably lead to a correct conclusion; it gives little attention to questions of how anatomical data are related to evolutionary theory or to phylogeny. For example, the shape of one human tooth, the lower first premolar, has been cited as proof that man never went through an apelike stage in the course of his evolution. As late as 1972 this datum was offered as evidence that man and his ancestors had been separated from the other primates for at least 35 million years. Since then late Pliocene hominid fossils have been found that are about 3.7 million years old, and their lower first premolars show moderately apelike characteristics. It was not the description of the tooth that was wrong but the conclusions drawn from it (not to mention the belief that it is reasonable to base major phylogenetic determinations on the anatomy of a single tooth).

Comparative anatomy nonetheless makes many valuable connections. For example, the bones of the human arm are much like those of an ape's arm but are very different from the comparable bones of a monkey. Monkey arm bones are very similar to those of many other primates and indeed to those of many other mammals; their form is basic to quadrupedal locomotion. In contrast, the form of human and ape arm bones is basic to the motions of climbing. This finding is a significant one, but it can lead to two quite opposite conclusions: (1) man and apes are related or (2) man and apes have followed a parallel course of evolution, that is, the structures of the arm evolved in the same way even though the two evolutionary lines had long since separated. Deciding between the two alternatives is made all the more difficult by the fact that the comparison is being made between two living animals, each of which has evolved to an unknown degree since its divergence from a common ancestry. Fortunately powerful new analytical tools have come into existence that are a great help in resolving such dilemmas. To these I shall return, under the heading of molecular anthropology; for the moment it is necessary only to say that when the patterns of primate biology being compared are functional ones, the fit between the conclusions drawn from comparative anatomy and those drawn from molecular anthropology is quite close.

Until a few decades ago the fossil record of the primates was poor and that of the hominids, including man, was even poorer. For example, when Sir Arthur Keith undertook to array the existing hominid fossils along their probable lines of descent some 50 years ago, he had to deal with only three genera in the Miocene epoch, and he was able to spread the five (at that time) hominid genera across a later Pliocene-to-Recent time interval of less than half a million years. (Between the Pliocene and the Recent epochs was the Pleistocene; to it and the Recent combined was allotted 200,000 years.) The five genera were *Homo erectus* (then represented only by specimens from Java named *Pithecanthropus*), Neanderthal man, Piltdown man (then still accepted as a valid genus named *Eoanthropus*), Rhodesian man (*Homo rhodesiensis*, a form no longer considered a distinct species) and finally the genus and species *Homo sapiens* (from which, as can be seen, Keith excluded the Neanderthals). Keith had Java man branching off from the main human stem in Miocene times and indicated the extinction of the line at the start of the Pleistocene. The Neanderthals, today classified as *Homo sapiens neanderthalensis*, he saw as branching off in the mid-Pliocene, shortly before the appearance of Rhodesian man and well before that of Piltdown man; he had all three genera extinct in Pleistocene times.

Keith's arrangement was marvelous in its simplicity: each fossil that had to be accounted for stood at the end of its own evolutionary branch, and the time of branching was deduced from its anatomy. (Piltdown man was a problem: its genuinely modern cranium put this faked specimen's branching point higher up on the tree than Rhodesian man's, but its genuinely nonhuman jawbone demanded that the branching be put in the Pliocene.) This kind of typological thinking died slowly with the discovery of many new hominid fossils and the rise of radioactive-isotope dating, which have extended the duration of the Pleistocene from about 200,000 years to about two million. Theodosius Dobzhansky's 1944 paper "On Species and Races of Living and Fossil Man" ushered in the new era and ended nearly a century of analysis that had been primarily typological.

Today there are hundreds of primate fossils. Many of them are accurately dated and more are being discovered every year. It is no longer even practical to list the individual specimens, which was the custom until a few years ago. Problems still remain with respect to the hominid fossil record, perhaps in part because human beings are obsessively curious about the details of their own ancestry. If any animal other than the human one were involved, the hominid fossil record over the past four million years would be considered adequate and even generous.

How is the evidence of those four million years to be read? To begin with, it can now be said with some certainty that hominids have walked upright for at least three million years. That is the age of a pelvis of the early hominid *Australopithecus* recently unearthed in the Afar region of Ethiopia by Donald C. Johanson of Case Western Reserve University. Prior to Johanson's find the best evidence of upright walking was a younger *Australopithecus* pelvis uncovered at Sterkfontein in South Africa. The two fossils are nearly identical. The inference is inescapable that bipedal locomotion is not just another human anatomical adaptation but the most fundamental one. The early bipeds all had small brains (average: 450 cubic centimeters).

Not much later, perhaps about 2.5 million years ago, the bipeds were making stone tools and hunting animals for food. By about two million years ago hominid craniums with a larger capacity appeared; by 1.5 million years ago *Homo erectus* was on the scene, the brains had doubled in size and the stone tools now include bifaces, tools that have been flaked on both sides. These bifaces belong to the core-tool industry known as the Acheulian. (The characteristic form was first recognized at a French Paleolithic site, St.-Acheul.) From about two million to one million years ago another kind of early biped was also present; its robust anatomy identifies it as a separate species of *Australopithecus*. It is readily distinguishable from the less robust bipeds by its massive jaw and molar teeth that are very large compared with the incisors.

This summary of the hominid fossil record is undoubtedly oversimplified, but I think the evidence supports the main outline. What difficulties there are arise mainly from the fragmentary nature of many of the fossils. For example, Johanson has found one skeleton in the Afar region complete enough to allow reconstruction of that hominid's general proportions. The reconstruction shows that it had relatively long arms, a fact that could not be determined from the hundreds of previously discovered fragmentary remains of *Australopithecus*.

Dating also causes problems. For ex-

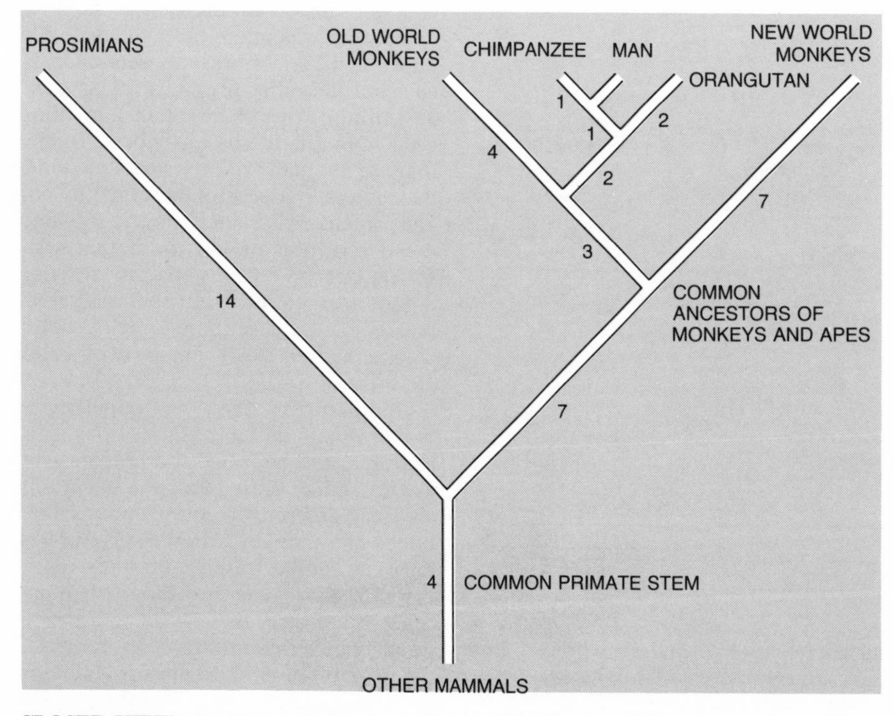

CLOSER VIEW of primate divergence is afforded by this schematic structure. The distance between man and the chimpanzee has a value of 1; this places both man and chimpanzee at a remove of 4 from the orangutan, and the orangutan and the Old World monkeys at a remove of 7 from the ancestor that both have in common with the New World monkeys. All anthropoids are at a remove of 7 from the ancestor they have in common with all prosimians (less advanced primates such as the lemurs) and at a remove of 11 from the most primitive primates.

ample, there are no radioactive-isotope dates for the hominid fossils found in South Africa. There is disagreement among specialists about the date of a particularly important marker layer of volcanic tuff in the East Turkana region of Kenya, where many important hominid fossils are currently being found. My response (or my bias) is to try to see the general order and to add complications only when they are absolutely inescapable.

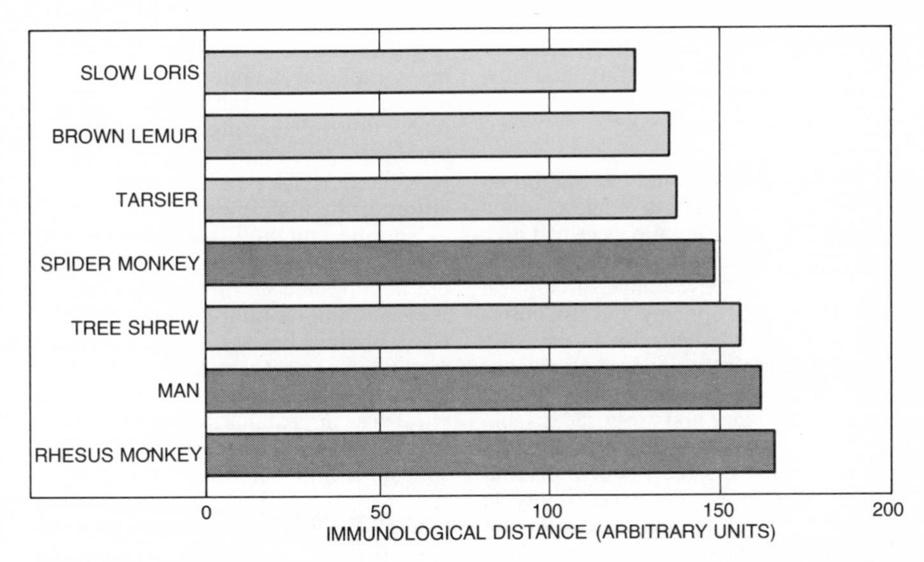

RATE OF EVOLUTION appears to be independent of the number of generations per unit time according to immunological-distance data. Bars show the distances, calculated by Vincent M. Sarich of the University of California at Berkeley, separating the carnivores from various primates. Man and the rhesus monkey were the most distant, respectively 162 and 166 units removed, although each human generation is five times longer than a rhesus generation. Four prosimians (*color*), also far shorter-lived than man, were even less distant from the carnivores.

The first conclusion I draw from this simplified picture is that upright walking evolved millions of years before a large brain, stone tools or other characteristics we think of as being human. If one accepts this conclusion, the problem of tracing human origins is primarily one of unearthing fossil evidence for that complex locomotor adaptation. How much time the adaptation required and what its intermediate stages may have been cannot be determined as long as the fossil leg bones are missing. The adaptation may have begun at any time from five to 10 million years ago. Fossil-bearing deposits of that age exist, and so all that is needed to clarify this aspect of human evolution is money for the search and a bit of luck.

The second conclusion I draw from my simplified outline is that stone tools and hunting long antedate the appearance of a large brain. Excavating in East Turkana, Glynn Isaac of the University of California at Berkeley and his colleagues uncovered a scatter of crude stone tools including both flakes and the cores that had yielded the flakes, and together with the tools were bits of animal bone. Unfortunately the creatures that deposited this material, which may be as much as 2.5 million years old, left no evidence of their own anatomy.

The East Turkana tools are very early but it is most unlikely that they are the earliest. For example, at Olduvai in neighboring Tanzania many of the stone tools from Bed I are unworked stones. They could not have been identified as tools except for the fact that they were found in a layer of volcanic ash otherwise free of stones, so that someone must have brought them there from somewhere else. In the absence of some similar circumstance the earliest stone tools are likely to go unrecognized.

My third conclusion stems both from the fossil record and from what is known about the anatomy of the human brain. As I have indicated, in my view large brains follow long after stone tools. Tools that are hard to make, such as those of the Acheulian industry, follow the earlier simple tools only after at least a million years have passed. It looks as if the successful way of life the earlier tools made possible acted in some kind of feedback relation with the evolution of the brain. What can be seen in the cortex of the human cerebrum mirrors this evolutionary success. Just as the proportions of the human hand, with its large and muscular thumb, reflect a selection for success in the use of tools, so does the anatomy of the human brain reflect a selection for success in manual skills.

Here a further point arises that is often forgotten. The only direct evidence for the importance of increasing brain size comes from the archaeological record. The hypothesis of a correlation between tool use and a large brain argues that the archaeological progression (from no stone tools to simple stone tools to tools of increasing refinement) is correlated with the doubling of hominid brain size. If the hypothesis is correct, the brain should not only have grown in size but also have increased in complexity. The fact remains that the fossil record contains no clues bearing on this neurological advance. Nevertheless, increasing brain size does seem to be correlated with the increasing complexity of stone tools over hundreds of thousands of years in a way that is not evident during the past 100,000 years of human evolution.

Students of the fossil record rely almost entirely on description, just as students of comparative anatomy do. When a fossil is discovered, the first requirement is to determine its geological context, its associations and its probable age. Once the fossil has been brought to the laboratory it is described and compared with similar fossils, and conclusions are drawn on the basis of the comparisons. The anatomical structures that are compared are complex ones and the work is arduous, so that a great deal of analysis remains to be completed on fossils first discovered many years ago. The method has other limitations. For example, teeth are traditionally compared in isolation one at a time. In nature, of course, upper and lower teeth interact where they meet. Comparisons that take this functional factor into account, as the anatomist W. E. Le Gros Clark has shown, give results quite different from those yielded by the traditional tooth-by-tooth method.

Comparative methods are further complicated by the simple fact that a face is full of teeth. The form of the face is related both to the teeth and to the chewing muscles; functional patterns of this kind are not well described by linear measurements. What is more, the descriptive tradition even sets limits on what is observed. For example, the lower jaw of the robust species of *Australopithecus* from East Africa has a very large ascending ramus, the part of the lower jaw that projects upward to hinge with the skull. At the top of the ascending ramus is what is called the mandibular condyle. As this species' lower jaw opened and its mandibular condyle moved forward, the teeth of its upper and lower jaws must have moved farther apart than they do in any other primate. When one considers all that has been written about the possible diet of *Australopithecus* and about this hominid's teeth, it is surprising to find a fact as fundamental as the size of its bite is not discussed.

The same jaw provides another example of the weakness of such descriptive systems. In the robust species of *Australopithecus* the inside of the ascending ramus has features unlike those found in any other primate. This fact is not mentioned in the formal descriptions because it is not traditional to study the inside of the ramus. I could give a number of other examples, but the point at issue is the same in all of them: there are no clearly defined rules that state how fossils should be compared or how anatomy should be understood.

Having sketched what the traditional disciplines have to suggest on the subject of man's evolution, we can now turn to the suggestions that stem from work in two relatively new disciplines: molecular anthropology and the observation of primate behavior in the wild. The first of these disciplines actually has a longer history than plate tectonics: whereas Wegener first proposed his theory in 1912, George H. F. Nuttall demonstrated that the biochemical classification of animals was a possibility in 1904. Nuttall's method was immunological. If blood serum from an animal is injected into an experimental animal, the experimental animal will manufacture antibodies against proteins in the foreign serum. If serum from the experimental animal is added to serum from a third animal, the antibodies will combine with similar proteins in that serum to form a precipitate. The stronger the precipitation reaction, the closer the relation of the first animal to the third.

Nuttall's method was successfully applied in a number of investigations, but he attracted no more disciples than Wegener did. Not until the past decade, when findings based on immunological methods were seen to agree with those based on the similarity of amino acid sequences in proteins and the similarity of nucleotide sequences in DNA, did the concept of molecular taxonomy gain acceptance. As with the radioactive-isotope methods for determining absolute dates, the new molecular methods are objective and quantitative; they yield the same results when the tests are conducted by different workers.

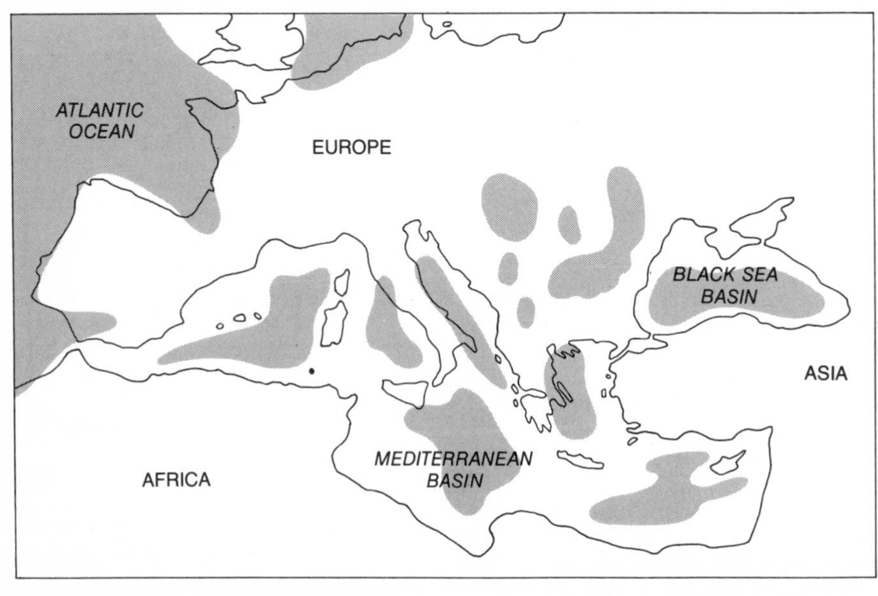

SEPARATION OF ASIA AND AFRICA, now joined by a narrow land bridge, was absolute some 20 million years ago (*top*), when the Tethys Sea reached from the Atlantic to the Persian Gulf. Later and until five million years ago (*bottom*) the Tethys was reduced to a network of lakes; the Old World primates were thus free at the time to move between the two continents.

The capacity of molecular taxonomy to define the relations among primates is perhaps the most important development in the study of human evolution over the past several decades. The great strength of the method is of course its objectivity. For example, data from the fossil record and from comparative anatomy have been cited to demonstrate that man's closest relative is variously the tarsier, certain monkeys, certain extinct apes, the chimpanzee or the gorilla, and to suggest that the time separating man from the last ancestor he shares with each of these candidates is variously from 50 million to four million years.

What do the data of molecular taxonomy show? The primary finding is that the molecular tests indicate little "distance" between man and the African apes. For example, when the distance separating the Old World monkeys from the New World monkeys is given a value of 1 and other distances are expressed as fractions of that value, then the distance between man and the Old World monkeys, as Vincent M. Sarich of the University of California at Berkeley has shown, is more than half a unit (.53 to .61). The distance between man and the Asian great ape the orangutan is about a quarter of a unit (.25 to .33) and the distance between man and the chimpanzee is about an eighth of a unit (.12 to .15).

The short distance between man and the African apes can be compared with similar distances among other related mammals. The relationship is about as close as that between horses and zebras and closer than that between dogs and foxes. Mary-Claire King and Allan C. Wilson of the University of California at Berkeley estimate (on the basis of comparisons between human and chimpanzee polypeptides, or protein chains) that man and the chimpanzee share more than 99 percent of their genetic material.

It might be thought such a wealth of new information about primate relationships would have been welcomed by students of human evolution. This has not been the case. The problem is that whereas the molecular data prove that man and the African apes are very closely related, the data appear to measure relationship and not time. It may be, however, that they do both. The overall picture seems clear: animals that are phylogenetically distant relatives are separated by large molecular distances and those that are close relatives are separated by small distances. This suggests that time and molecular distance are correlated. Unfortunately when it comes to the primates, the molecular distance between the New World and Old World monkeys is much too small to fit in with conventional phylogeny. What is worse, the distance between man and the African apes is startlingly less than convention demands. I suspect

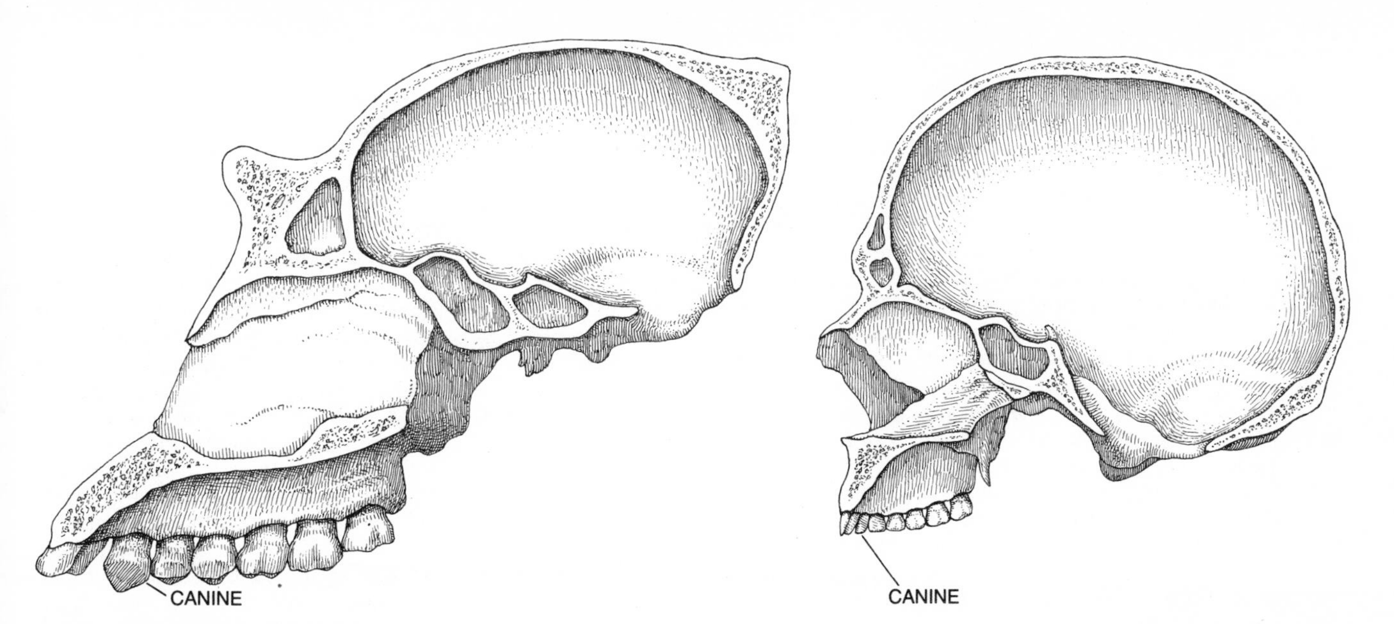

SKULLS OF MAN AND THE GORILLA, seen here in centerline section, have in common upper canine teeth that are much reduced in size. The gorilla is a female; male gorillas bare their very large canines when threatening to fight. The suggestion by Charles Darwin that man's use of weapons relieved him of the need for large canines seems to be supported by fossil evidence: the oldest human canines known are quite small compared with the canines of male African apes. This implies man's use of weapons for hundreds of millenniums.

that if molecular anthropology had shown man and apes to be very far apart, the concept of a correlation between genetic difference and time would have been accepted without debate.

The validity of a molecular clock is being argued at present, but I believe the problems will be worked out over the next few years. The chemical techniques are being improved and the fund of relevant information is being enlarged by work in many laboratories. Meanwhile the fossil evidence makes it highly unlikely that the ape and human lines separated less than five million years ago, and the molecular evidence makes it highly unlikely that they separated more than 10 million years ago. I have friends and colleagues who violently attack both dates, and they may be right! I am impressed by the degree of emotion that still surrounds the study of human evolution.

Studies of monkeys and apes under natural conditions have increased in number over the past few years. It is noteworthy in this connection that the heyday of evolutionary speculation was in the 19th century and that almost all the primate field studies began after 1960. The brutish, stooped Neanderthal and the monogamous chimpanzee have both proved to be products of the 19th-century imagination. Perhaps the most pertinent revision of preconceived views has to do with locomotion. All the traditional theories of human origins carefully considered how it was that a tree-dwelling ancestor became a ground-dwelling upright walker. The field studies have shown that our closest primate relatives, the African apes, are primarily ground dwellers. Moreover, their locomotor patterns suggest that the ancestor we share in common with them, however long ago, was also a ground dweller.

In the quadrupedal locomotion of most primates the hand and the foot are both placed flat on the ground; the animal cannot carry anything in them and move at the same time. Gorillas and chimpanzees (and the men who play some of the forward positions in American football), however, have developed a form of locomotion called knuckle walking that enables the apes (if not the football players) to walk normally as they carry objects between their fingers and their palm. If knuckle walking is an ancient trait, it neatly gets around the problem of how the handling and using of objects could have become a common habit. Of all living mammals except man the knuckle-walking chimpanzee is the most habitual user of objects. As Jane Goodall and her colleagues at the Gombe Stream Research Centre in Tanzania add to their observations year after year, the record of the number of objects handled by chimpanzees and the number of ways they are employed steadily increases. The chimpanzees use sticks for bluffing and attack, for poking, teasing and exploring. They use twigs and blades of grass to collect termites and ants. They use leaves to clean themselves. They use stones to crack nuts and also throw stones with moderate accuracy.

Our incredulity dies hard. When Peking man (now classified as *Homo erectus*) was first found, he was declared to be far too primitive to have made the stone tools found in association with his remains. The next unjustified victim of incredulity was *Australopithecus;* surely, the consensus had it, no one with such a small brain could have made tools. Even today many believe only one form of early biped, the form ancestral to man, could have made tools. Chimpanzee behavior is therefore enlightening: it shows that a typical ape is able to use objects in far greater variety and with greater effectiveness than anyone had suspected. There is no longer any reason not to suppose all the early bipeds also used objects, and probably used them far more than chimpanzees do, from a time far earlier than the time when stone tools first appear in the archaeological record.

Darwin suggested that the reason men had small canine teeth and the gorilla had huge ones was that man's possession of weapons had eliminated the need for fangs. It is clear that the large canine of the male gorilla has nothing to do with efficient chewing; the canine of the female gorilla is small but she is as well nourished as the male. Is the male canine part of an adaptation for bluffing and fighting? Before such an anatomical feature could have been reduced in the course of evolution its offensive function would have had to have been transferred to some other structure or mechanism. On this view the evolution of small human canine teeth would, as Darwin surmised, have depended on the use of weapons. The chimpanzee field studies, with their evidence for the frequent use of objects, support Darwin's interpretation. Sticks are seldom fossilized and unworked stones can rarely be proved to be artifacts, but teeth are the commonest of all hominid fossils and the earliest bipeds already had small canines. They had probably been using tools for many hundreds of thousands of years.

Not all behavioral information comes from studies in the field. Consider speech. The nonhuman primates cannot learn to speak even though great efforts have been made in the laboratory to teach them to do so. The recent remarkable successes in teaching apes how to communicate by symbols have been achieved in ways other than verbal ones. There is a lesson here, since human beings learn to speak with the greatest of ease.

The sounds made by monkeys primarily convey emotions and are controlled by brain systems more primitive than the cerebral cortex; removal of the cortex does not affect the production of sounds. In man the cortex of the dominant side of the brain is very important in speech. Speech is of course the form of behavior that more than any other differentiates man from other animals. Yet in spite of many ingenious attempts at investigation the origins of human speech remain a mystery. There is no clue to its presence or absence to be found in the fossils.

The archaeological record, however, does offer clues. What we see in the last 40,000 years of prehistory may have been triggered by the development of speech as we know it today. This is to say that although man was surely not mute for most of his development, an increased capacity for verbal communication may have been the ability that led to the extraordinary spread of modern man, *Homo sapiens sapiens*.

For most of the past million years the progress of human evolution, both biological and technological, was slow. Traditions of stone-tool manufacture, as reflected in the rise of successive stone-tool industries, persisted for hundreds of thousands of years with little change. Then came the great acceleration of about 40,000 years ago. Men who were anatomically modern now dominated the scene. Primitive forms of man disappeared; there are not enough fossils to make it possible to decide whether the disappearance was by evolution, hybridization or extinction. Then, in far less than 1 percent of the time that bipeds are present in the fossil record, came a technological revolution. Its fruits included entirely new and complex tools and weapons, the construction of shelters, the invention of boats, the addition of fish and shellfish to the human diet, deep-water voyages (to Australia, for example), the peopling of the Arctic, the migration to the Americas and the proliferation of a lively variety of arts and a wide range of personal adornment.

The rate of change continued to accelerate. Agriculture and animal husbandry appeared at roughly the same time around the globe. Technological progress, the mastery of new materials (such as metals) and new energy sources (such as wind and water power) led in an amazingly short time to the Industrial Revolution and the world of today. The acceleration of human history cannot be better illustrated than by comparing the changes of the past 10,000 years with those of the previous four million.

Language, that marriage of speech and cognitive abilities, may well have been the critical new factor that provided a biological base for the acceleration of history. Just as upright walking and toolmaking were the unique adaptation of the earlier phases of human evolution, so was the physiological capacity for speech the biological base for the later stages. Without this remarkably effective mode of communication man's technological advance would perforce have been slow and limited. Given an open system of communication rapid change becomes possible and social systems can grow in complexity. Human social systems are all mediated by language; perhaps this is why there are no forms of behavior among the nonhuman primates that correspond to religion, politics or even economics.

If all this seems too pat, I should remind the reader that some of the oldest and most troublesome questions about human evolution remain unanswered. Looking to the future, I expect that molecular biology will determine the relationships between man and the other living primates and the times of their mutual divergence more accurately than any other discipline can. But there will still be other major problems, particularly in determining the rates of evolution. As in the past, the present proponents of various hypotheses may be wrong on the very points on which they are surest they are right.

At this stage, then, it is probably wise to entertain more than one hypothesis and to state opinions in terms of the odds in favor of their being right rather than presenting them as conclusions. On this basis I would guess from the present evidence that the odds are 100 to one (in favor) that man and the African apes do in fact form a closely related group. I would also guess that a very recent separation of man and the apes, say five to six million years ago, was not nearly as probable; there my odds are only two to one in favor.

Perhaps by presenting opinions in this way we might demonstrate that all views of human evolution are built on seeming facts that vary widely in their degree of reliability. For example, if it is accepted that man is particularly close in his relationship to the African apes, it does not necessarily follow that man and the apes separated in Africa. At the time when the ape and human lines separated there were apes in the Near East and India; man may be descended from the apes of those areas. Perhaps the reason there are no longer any apes in India is that those apes evolved into men. Both the African and the non-African theories of the evolution of the earliest upright walkers are reasonable; only the discovery of more fossils will determine which theory is correct.

IX

Adaptation

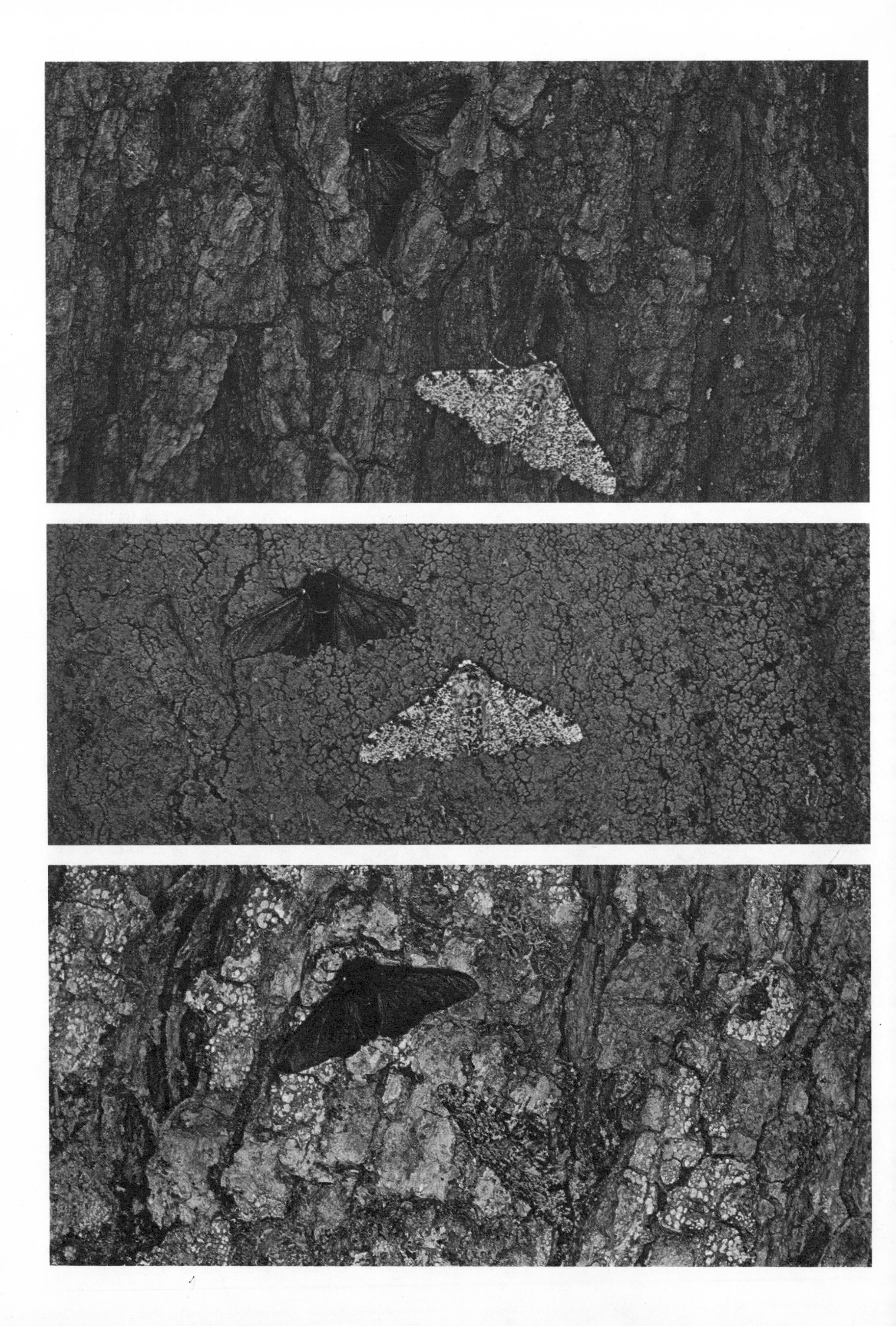

Adaptation

BY RICHARD C. LEWONTIN

The manifest fit between organisms and their environment is a major outcome of evolution. Yet natural selection does not lead inevitably to adaptation; indeed, it is sometimes hard to define an adaptation

The theory about the history of life that is now generally accepted, the Darwinian theory of evolution by natural selection, is meant to explain two different aspects of the appearance of the living world: diversity and fitness. There are on the order of two million species now living, and since at least 99.9 percent of the species that have ever lived are now extinct, the most conservative guess would be that two billion species have made their appearance on the earth since the beginning of the Cambrian period 600 million years ago. Where did they all come from? By the time Darwin published *On the Origin of Species* in 1859 it was widely (if not universally) held that species had evolved from one another, but no plausible mechanism for such evolution had been proposed. Darwin's solution to the problem was that small heritable variations among individuals within a species become the basis of large differences between species. Different forms survive and reproduce at different rates depending on their environment, and such differential reproduction results in the slow change of a population over a period of time and the eventual replacement of one common form by another. Different populations of the same species then diverge from one another if they occupy different habitats, and eventually they may become distinct species.

Life forms are more than simply multiple and diverse, however. Organisms fit remarkably well into the external world in which they live. They have morphologies, physiologies and behaviors that appear to have been carefully and artfully designed to enable each organism to appropriate the world around it for its own life.

It was the marvelous fit of organisms to the environment, much more than the great diversity of forms, that was the chief evidence of a Supreme Designer. Darwin realized that if a naturalistic theory of evolution was to be successful, it would have to explain the apparent perfection of organisms and not simply their variation. At the very beginning of the *Origin of Species* he wrote: "In considering the Origin of Species, it is quite conceivable that a naturalist . . . might come to the conclusion that each species . . . had descended, like varieties, from other species. Nevertheless, such a conclusion, even if well founded, would be unsatisfactory, until it could be shown how the innumerable species inhabiting this world have been modified, so as to acquire that perfection of structure and coadaptation which most justly excites our admiration." Moreover, Darwin knew that "organs of extreme perfection and complication" were a critical test case for his theory, and he took them up in a section of the chapter on "Difficulties of the Theory." He wrote: "To suppose that the eye, with all its inimitable contrivances for adjusting the focus to different distances, for admitting different amounts of light, and for the correction of spherical and chromatic aberration, could have been formed by natural selection, seems, I freely confess, absurd in the highest degree."

These "organs of extreme perfection" were only the most extreme case of a more general phenomenon: adaptation. Darwin's theory of evolution by natural selection was meant to solve both the problem of the origin of diversity and the problem of the origin of adaptation at one stroke. Perfect organs were a difficulty of the theory not in that natural selection could not account for them but rather in that they were its most rigorous test, since on the face of it they seemed the best intuitive demonstration that a divine artificer was at work.

The modern view of adaptation is that the external world sets certain "problems" that organisms need to "solve," and that evolution by means of natural selection is the mechanism for creating these solutions. Adaptation is the process of evolutionary change by which the organism provides a better and better "solution" to the "problem," and the end result is the state of being adapted. In the course of the evolution of birds from reptiles there was a successive alteration of the bones, the muscles and the skin of the forelimb to give rise to a wing; an increase in the size of the breastbone to provide an anchor for the wing muscles; a general restructuring of bones to make them very light but strong, and the development of feathers to provide both aerodynamic elements and lightweight insulation. This wholesale reconstruction of a reptile to make a bird is considered a process of major adaptation by which birds solved the problem of flight. Yet there is no end to adaptation. Having adapted to flight, some birds reversed the process: the penguins adapted to marine life by changing their wings into flippers and their feathers into a waterproof covering, thus solving the problem of aquatic existence.

The concept of adaptation implies a preexisting world that poses a problem to which an adaptation is the solution. A key is adapted to a lock by cutting and filing it; an electrical appliance is adapted to a different voltage by a transform-

ADAPTATION is exemplified by "industrial melanism" in the peppered moth (*Biston betularia*). Air pollution kills the lichens that would normally colonize the bark of tree trunks. On the dark, lichenless bark of an oak tree near Liverpool in England the melanic (*black*) form is better adapted: it is better camouflaged against predation by birds than the light, peppered wild type (*top photograph on opposite page*), which it largely replaced through natural selection in industrial areas of England in the late 19th century. Now air quality is improving. On a nearby beech tree colonized by algae and the lichen *Lecanora conizaeoides*, which is itself particularly well adapted to low levels of pollution, the two forms of the moth are equally conspicuous (*middle*). On the lichened bark of an oak tree in rural Wales the wild type is almost invisible (*bottom*), and in such areas it predominates. The photographs were made by J. A. Bishop of the University of Liverpool and Laurence M. Cook of the University of Manchester.

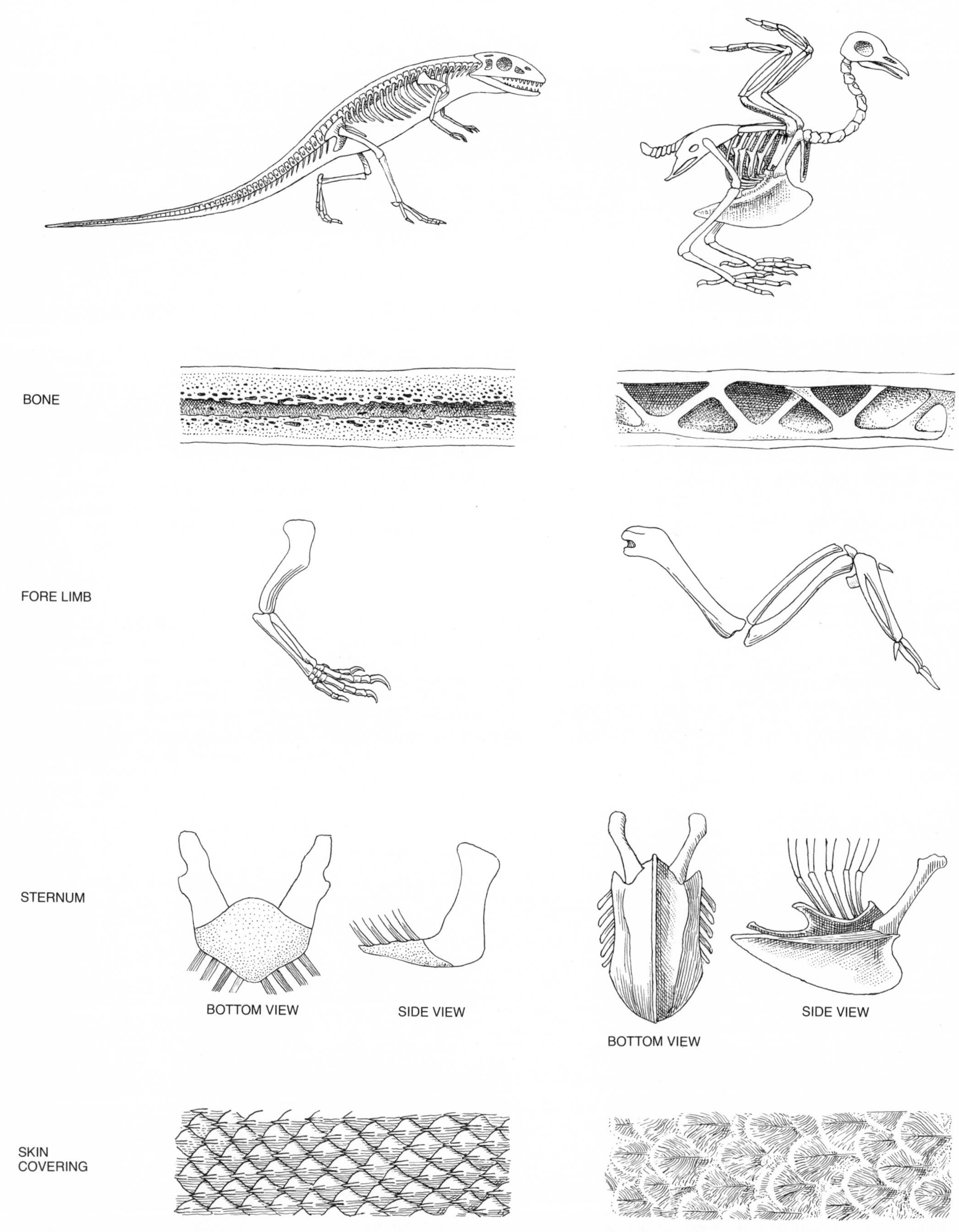

EVOLUTION OF BIRDS from reptiles can be considered a process of adaptation by which birds "solved" the "problem" of flight. At the top of the illustration the skeleton of a modern pigeon (*right*) is compared with that of an early reptile: a thecodont, a Triassic ancestor of dinosaurs and birds. Various reptile features were modified to become structures specialized for flight. Heavy, dense bone was restructured to become lighter but strong; the forelimb was lengthened (and its muscles and skin covering were changed) to become a wing; the reptilian sternum, or breastbone, was enlarged and deepened to anchor the wing muscles (even in *Archaeopteryx,* the Jurassic transition form between reptiles and birds whose sternum is pictured here, the sternum was small and shallow); scales developed into feathers.

er. Although the physical world certainly predated the biological one, there are certain grave difficulties for evolutionary theory in defining that world for the process of adaptation. It is the difficulty of defining the "ecological niche." The ecological niche is a multidimensional description of the total environment and way of life of an organism. Its description includes physical factors, such as temperature and moisture; biological factors, such as the nature and quantity of food sources and of predators, and factors of the behavior of the organism itself, such as its social organization, its pattern of movement and its daily and seasonal activity cycles.

The first difficulty is that if evolution is described as the process of adaptation of organisms to niches, then the niches must exist before the species that are to fit them. That is, there must be empty niches waiting to be filled by the evolution of new species. In the absence of organisms in actual relation to the environment, however, there is an infinity of ways the world can be broken up into arbitrary niches. It is trivially easy to describe "niches" that are unoccupied. For example, no organism makes a living by laying eggs, crawling along the surface of the ground, eating grass and living for several years. That is, there are no grass-eating snakes, even though snakes live in the grass. Nor are there any warm-blooded, egg-laying animals that eat the mature leaves of trees, even though birds inhabit trees. Given any description of an ecological niche occupied by an actual organism, one can create an infinity of descriptions of unoccupied niches simply by adding another arbitrary specification. Unless there is some preferred or natural way to subdivide the world into niches the concept loses all predictive and explanatory value.

A second difficulty with the specification of empty niches to which organisms adapt is that it leaves out of account the role of the organism itself in creating the niche. Organisms do not experience environments passively; they create and define the environment in which they live. Trees remake the soil in which they grow by dropping leaves and putting down roots. Grazing animals change the species composition of herbs on which they feed by cropping, by dropping manure and by physically disturbing the ground. There is a constant interplay of the organism and the environment, so that although natural selection may be adapting the organism to a particular set of environmental circumstances, the evolution of the organism itself changes those circumstances. Finally, organisms themselves determine which external factors will be part of their niche by their own activities. By building a nest the phoebe makes the availability of dried grass an important part of its niche, at the same time making the nest itself a component of the niche.

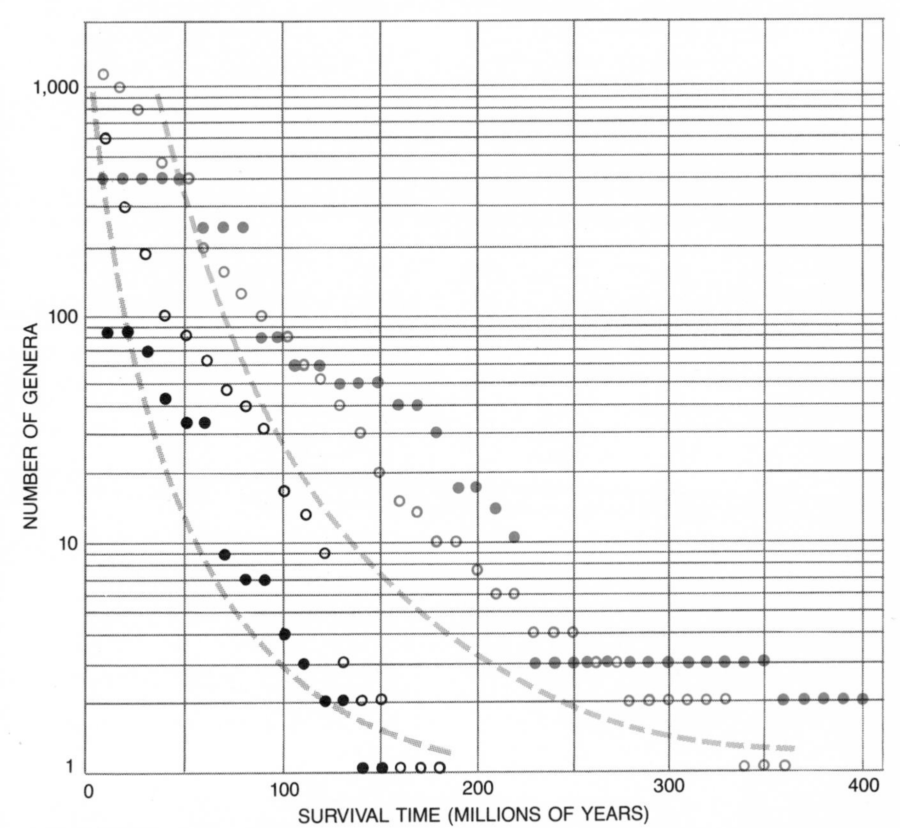

EXTINCTION RATES in many evolutionary lines suggest that natural selection does not necessarily improve adaptation. The data, from Leigh Van Valen of the University of Chicago, show the duration of survival of a number of living (*solid dots*) and extinct (*open circles*) genera of Echinoidea (*black*) and Pelecypoda (*color*), two classes of marine invertebrates. If natural selection truly fitted organisms to environments, the points should fall along concave curves (*broken-line curves*) indicating a lower probability of extinction for long-lived genera. Actually, points fall along rather straight lines, indicating constant rate of extinction for each group.

If ecological niches can be specified only by the organisms that occupy them, evolution cannot be described as a process of adaptation because all organisms are already adapted. Then what is happening in evolution? One solution to this paradox is the Red Queen hypothesis, named by Leigh Van Valen of the University of Chicago for the character in *Through the Looking Glass* who had to keep running just to stay in the same place. Van Valen's theory is that the environment is constantly decaying with respect to existing organisms, so that natural selection operates essentially to enable the organisms to maintain their state of adaptation rather than to improve it. Evidence for the Red Queen hypothesis comes from an examination of extinction rates in a large number of evolutionary lines. If natural selection were actually improving the fit of organisms to their environments, then we might expect the probability that a species will become extinct in the next time period to be less for species that have already been in existence for a long time, since the long-lived species are presumably the ones that have been improved by natural selection. The data show, however, that the probability of extinction of a species appears to be a constant, characteristic of the group to which it belongs but independent of whether the species has been in existence for a long time or a short one. In other words, natural selection over the long run does not seem to improve a species' chance of survival but simply enables it to "track," or keep up with, the constantly changing environment.

The Red Queen hypothesis also accounts for extinction (and for the occasional dramatic increases in the abundance and range of species). For a species to remain in existence in the face of a constantly changing environment it must have sufficient heritable variation of the right kind to change adaptively. For example, as a region becomes drier because of progressive changes in rainfall patterns, plants may respond by evolving a deeper root system or a thicker cuticle on the leaves, but only if their gene pool contains genetic variation for root length or cuticle thickness, and successfully only if there is enough genetic variation so that the species can change as fast as the environment. If the genetic variation is inadequate, the species will become extinct. The genetic resources of a species are finite, and eventually the environment will change so rapidly that the species is sure to become extinct.

The theory of environmental tracking seems at first to solve the problem of adaptation and the ecological niche. Whereas in a barren world there is no clear way to divide the environment into preexisting niches, in a world already occupied by many organisms the terms of the problem change. Niches are already defined by organisms. Small changes in the environment mean small changes in the conditions of life of those organisms, so that the new niches to which they must evolve are in a sense very close to the old ones in the multidimensional niche space. Moreover, the organisms that will occupy these slightly changed niches must themselves come from the previously existing niches, so that the kinds of species that can evolve are stringently limited to ones that are extremely similar to their immediate ancestors. This in turn guarantees that the changes induced in the environment by the changed organism will also be small and continuous in niche space. The picture of adaptation that emerges is the very slow movement of the niche through niche space, accompanied by a slowly changing species, always slightly behind, slightly ill-adapted, eventually becoming extinct as it fails to keep up with the changing environment because it runs out of genetic variation on which natural selection can operate. In this view species form when two populations of the same species track environments that diverge from each other over a period of time.

The problem with the theory of environmental tracking is that it does not predict or explain what is most dramatic in evolution: the immense diversification of organisms that has accompanied, for example, the occupation of the land from the water or of the air from the land. Why did warm-blooded animals arise at a time when cold-blooded animals were still plentiful and come to coexist with them? The appearance of entirely new life forms, of ways of making a living, is equivalent to the occupation of a previously barren world and brings us back to the preexistent empty niche waiting to be filled. Clearly there have been in the past ways of making a living that were unexploited and were then "discovered" or "created" by existing organisms. There is no way to explain and predict such evolutionary adaptations unless a priori niches can be described on the basis of some physical principles before organisms come to occupy them.

That is not easy to do, as is indicated by an experiment in just such a priori predictions that has been carried out by probes to Mars and Venus designed to detect life. The instruments are designed to detect life by detecting growth in nutrient solutions, and the solutions are prepared in accordance with knowledge of terrestrial microorganisms, so that the probes will detect only organisms whose ecological niches are like those on the earth. If Martian and Venusian life partition the environment in totally unexpected ways, they will remain unrecorded. What the designers of those instruments never dreamed of was that the reverse might happen: that the nature of the physical environment on Mars might be such that when it was provided with a terrestrial ecological niche, inorganic reactions might have a lifelike appearance. Yet that may be exactly what happened. When the Martian soil was dropped into the nutrient broth on the lander, there was a rapid production of carbon dioxide and then—nothing. Either an extraordinary kind of life began to grow much more rapidly than any terrestrial microorganism and then was poisoned by its own activity in a strange environment, or else the Martian soil is such that its contact with nutrient broths results in totally unexpected catalytic processes. In either case the Mars life-detection experiment has foundered on the problem of defining ecological niches without organisms.

Much of evolutionary biology is the working out of an adaptationist program. Evolutionary biologists assume that each aspect of an organism's morphology, physiology and behavior has been molded by natural selection as a solution to a problem posed by the

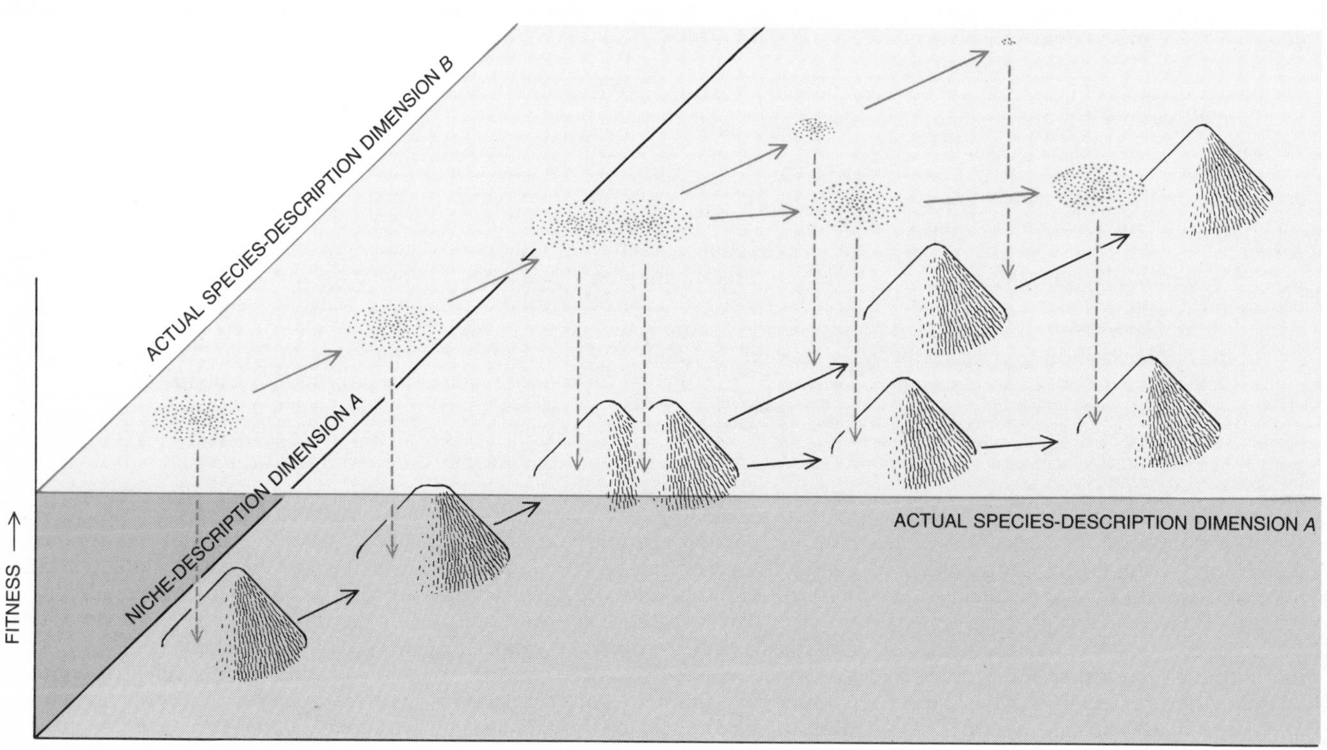

SPECIES TRACK ENVIRONMENT through niche space, according to one view of adaptation. The niche, visualized as an "adaptive peak," keeps changing (moving to the right); a slowly changing species population (*colored dots*) just manages to keep up with the niche, always a bit short of the peak. As the environment changes, the single peak becomes two distinct peaks, and two populations diverge to form distinct species. One species cannot keep up with its rapidly changing environment, becomes less fit (lags farther behind changing peak) and extinct. Here niche space and actual-species space have only two dimensions; both of them are actually multidimensional.

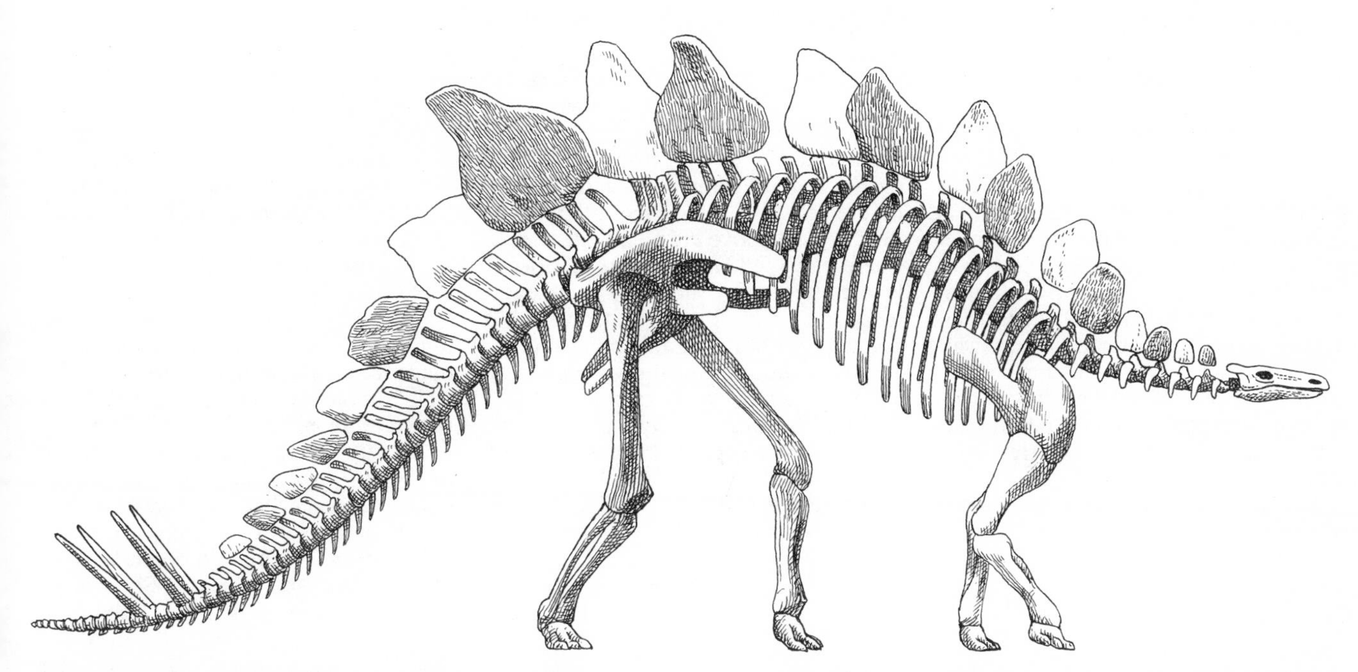

STEGOSAURUS, a large herbivorous dinosaur of the Jurassic period, had an array of bony plates along its back. Were they solutions to the problem of defense, courtship recognition or heat regulation? An engineering analysis reveals features characteristic of heat regulators: porous structure (suggesting a rich blood supply), particularly large plates over the massive part of the body, staggered arrangement along the midline, a constriction near the base and so on. This skeleton in the American Museum of Natural History is 18 feet long.

environment. The role of the evolutionary biologist is then to construct a plausible argument about how each part functions as an adaptive device. For example, functional anatomists study the structure of animal limbs and analyze their motions by time-lapse photography, comparing the action and the structure of the locomotor apparatus in different animals. Their interest is not, however, merely descriptive. Their work is informed by the adaptationist program, and their aim is to explain particular anatomical features by showing that they are well suited to the function they perform. Evolutionary ethologists and sociobiologists carry the adaptationist program into the realm of animal behavior, providing an adaptive explanation for differences among species in courting pattern, group size, aggressiveness, feeding behavior and so on. In each case they assume, like the functional anatomist, that the behavior is adaptive and that the goal of their analysis is to reveal the particular adaptation.

The dissection of an organism into parts, each of which is regarded as a specific adaptation, requires two sets of a priori decisions. First one must decide on the appropriate way to divide the organism and then one must describe what problem each part solves. This amounts to creating descriptions of the organism and of the environment and then relating the descriptions by functional statements; one can either start with the problems and try to infer which aspect of the organism is the solution or start with the organism and then ascribe adaptive functions to each part.

For example, for individuals of the same species to recognize each other at mating time is a problem, since mistakes about species mean time, energy and gametes wasted in courtship and mating without the production of viable offspring; species traits such as distinctive color markings, special courtship behavior, unique mating calls, odors and restricted time and place of activity can be considered specific adaptations for the proper recognition of potential mates. On the other hand, the large, leaf-shaped bony plates along the back of the dinosaur *Stegosaurus* constitute a specific characteristic for which an adaptive function needs to be inferred. They have been variously explained as solutions to the problem of defense (by making the animal appear to be larger or by interfering directly with the predator's attack), the problem of recognition in courtship and the problem of temperature regulation (by serving as cooling fins).

The same problems that arose in deciding on a proper description of the ecological niche without the organism arise when one tries to describe the organism itself. Is the leg a unit in evolution, so that the adaptive function of the leg can be inferred? If so, what about a part of the leg, say the foot, or a single toe, or one bone of a toe? The evolution of the human chin is an instructive example. Human morphological evolution can be generally described as a "neotenic" progression. That is, human infants and adults resemble the fetal and young forms of apes more than they resemble adult apes; it is as if human beings are born at an earlier stage of physical development than apes and do not mature as far along the apes' development path. For example, the relative proportion of skull size to body size is about the same in newborn apes and human beings, whereas adult apes have much larger bodies in relation to their heads than we do; in effect their bodies "go further."

The exception to the rule of human neoteny is the chin, which grows relatively larger in human beings, whereas both infant and adult apes are chinless. Attempts to explain the human chin as a specific adaptation selected to grow larger failed to be convincing. Finally it was realized that in an evolutionary sense the chin does not exist! There are two growth fields in the lower jaw: the dentary field, which is the bony structure of the jaw, and the alveolar field, in which the teeth are set. Both the dentary and the alveolar fields do show neoteny. They have both become smaller in the human evolutionary line. The alveolar field has shrunk somewhat faster than the dentary one, however, with the result that a "chin" appears as a pure consequence of the relative regression rates of the two growth fields. With the recognition that the chin is a mental construct rather than a unit in evolution the problem of its adaptive explanation disappears. (Of course, we may go on to ask why the dentary and alveolar growth fields have regressed at different rates in evolution, and then provide an adaptive explanation for that phenomenon.)

Sometimes even the correct topology of description is unknown. The brain is divided into anatomical divisions corresponding to certain separable nervous functions that can be localized, but memory is not one of those functions. The memory of specific events seems to be stored diffusely over large regions of the cerebrum rather than being localized microscopically. As one moves from anatomy to behavior the problem of a correct description becomes more acute and the opportunities to introduce arbitrary constructs as if they were evolutionary traits multiply. Animal behavior is described in terms of aggression, division of labor, warfare, dominance, slave-making, cooperation—and yet each of these is a category that is taken directly from human social experience and is transferred to animals.

The decision as to which problem is solved by each trait of an organism is equally difficult. Every trait is involved in a variety of functions, and yet one would not want to say that the character is an adaptation for all of them. The green turtle *Chelonia mydas* is a large marine turtle of the tropical Pacific. Once a year the females drag themselves up the beach with their front flippers to the dry sand above the high-water mark. There they spend many hours laboriously digging a deep hole for their eggs, using their hind flippers as trowels. No one who has watched this painful process would describe the turtles' flippers as adaptations for land locomotion and digging; the animals move on land and dig with their flippers because nothing better is available. Conversely, even if a trait seems clearly adaptive, it cannot be assumed that the species would suffer in its absence. The fur of a polar bear is an adaptation for temperature regulation, and a hairless polar bear would certainly freeze to death. The color of a polar bear's fur is another matter. Although it may be an adaptation for camouflage, it is by no means certain that the polar bear would become extinct or even less numerous if it were brown. Adaptations are not necessary conditions of the existence of the species.

For extinct species the problem of judging the adaptive status of a trait is made more difficult because both the trait and its function must be reconstructed. In principle there is no way to be sure whether the dorsal plates of *Stegosaurus* were heat-regulation devices, a defense mechanism, a sexual recognition sign or all these things. Even in living species where experiments can be carried out a doubt remains. Some modern lizards have a brightly colored dewlap under the jaw. The dewlap may be a warning sign, a sexual attractant or a species-recognition signal. Experiments removing or altering the dewlap could decide, in principle, how it functions. That is a different question from its status as an adaptation, however, since the assertion of adaptation implies a historical argument about natural selection as the cause of its establishment. The large dorsal plates of *Stegosaurus* may have evolved because individuals with slightly larger plates were better able to gather food in the heat of the day than other individuals. If, when the plates reached a certain size, they incidentally frightened off predators, they would be a "preadaptation" for defense. The distinction between the primary adaptation for which a trait evolved and incidental functions it may have come to have cannot be made without the reconstruction of the forces of natural selection during the actual evolution of the species.

The current procedure for judging the adaptation of traits is an engineering analysis of the organism and its environment. The biologist is in the position of an archaeologist who uncovers a machine without any written record and attempts to reconstruct not only its operation but also its purpose. The hypothesis that the dorsal plates of *Stegosaurus* were a heat-regulation device is based on the fact that the plates were porous and probably had a large supply of blood vessels, on their alternate placement to the left and right of the midline (suggesting cooling fins), on their large size over the most massive part of the body and on the constriction near their base, where they are closest to the heat source and would be inefficient heat radiators.

Ideally the engineering analysis can be quantitative as well as qualitative and so provide a more rigorous test of the

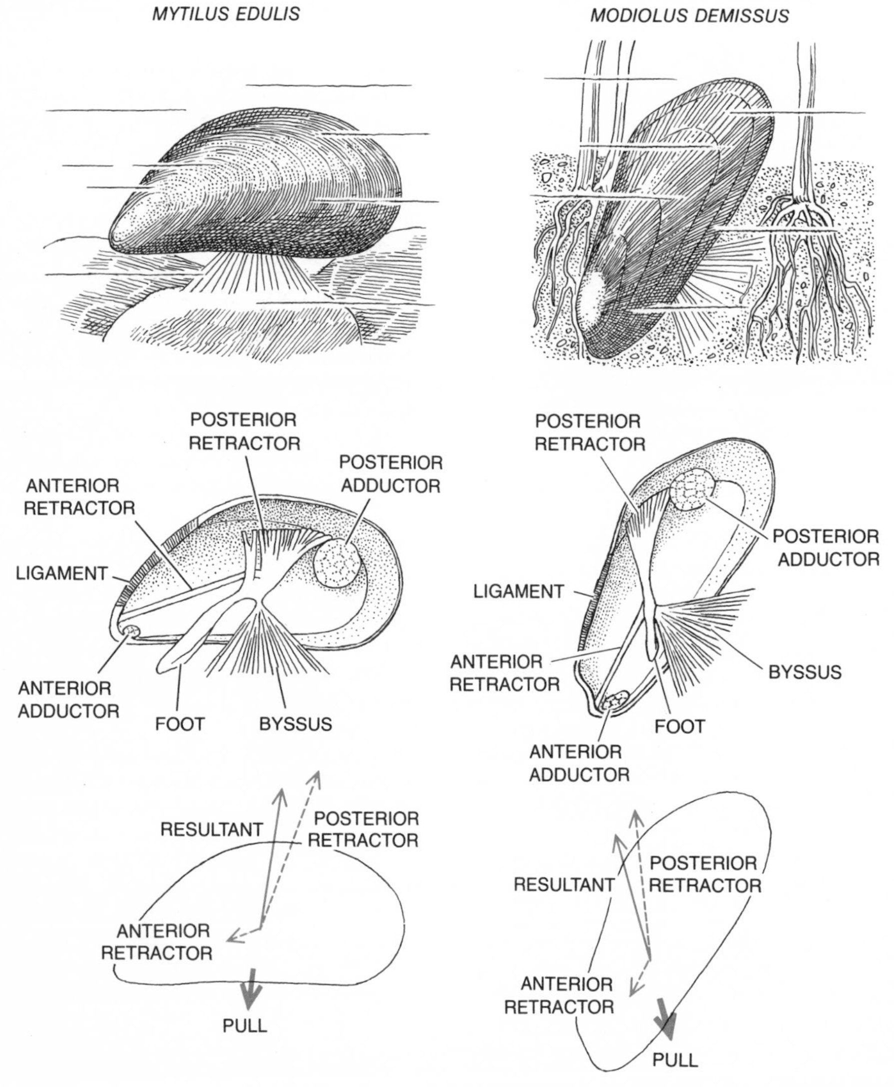

FUNCTIONAL ANALYSIS indicates how the shape and musculature of two species of mussels are adapted to their particular environments. *Mytilus edulis* (*left*) attaches itself to rocks by means of its byssus, a beardlike group of threads (*top*). Its ventral, or lower, edge is flattened; the anterior and posterior retractor muscles are positioned (*middle*) so that their resultant force pulls the bottom of the shell squarely down to the substratum (*bottom*). *Modiolus demissus* (*right*) attaches itself to debris in marshes. Its ventral edge is sharply angled to facilitate penetration of the substratum; its retractor muscles are positioned to pull its anterior end down into the marsh. The analysis was done by Steven M. Stanley of Johns Hopkins University.

adaptive hypothesis. Egbert G. Leigh, Jr., of the Smithsonian Tropical Research Institute posed the question of the ideal shape of a sponge on the assumption that feeding efficiency is the problem to be solved. A sponge's food is suspended in water and the organism feeds by passing water along its cell surfaces. Once water is processed by the sponge it should be ejected as far as possible from the organism so that the new water taken in is rich in food particles. By an application of simple hydrodynamic principles Leigh was able to show that the actual shape of sponges is maximally efficient. Of course, sponges differ from one another in the details of their shape, so that a finer adjustment of the argument would be needed to explain the differences among species. Moreover, one cannot be sure that feeding efficiency is the only problem to be solved by shape. If the optimal shape for feeding had turned out to be one with many finely divided branches and protuberances rather than the compact shape observed, it might have been argued that the shape was a compromise between the optimal adaptation for feeding and the greatest resistance to predation by small browsing fishes.

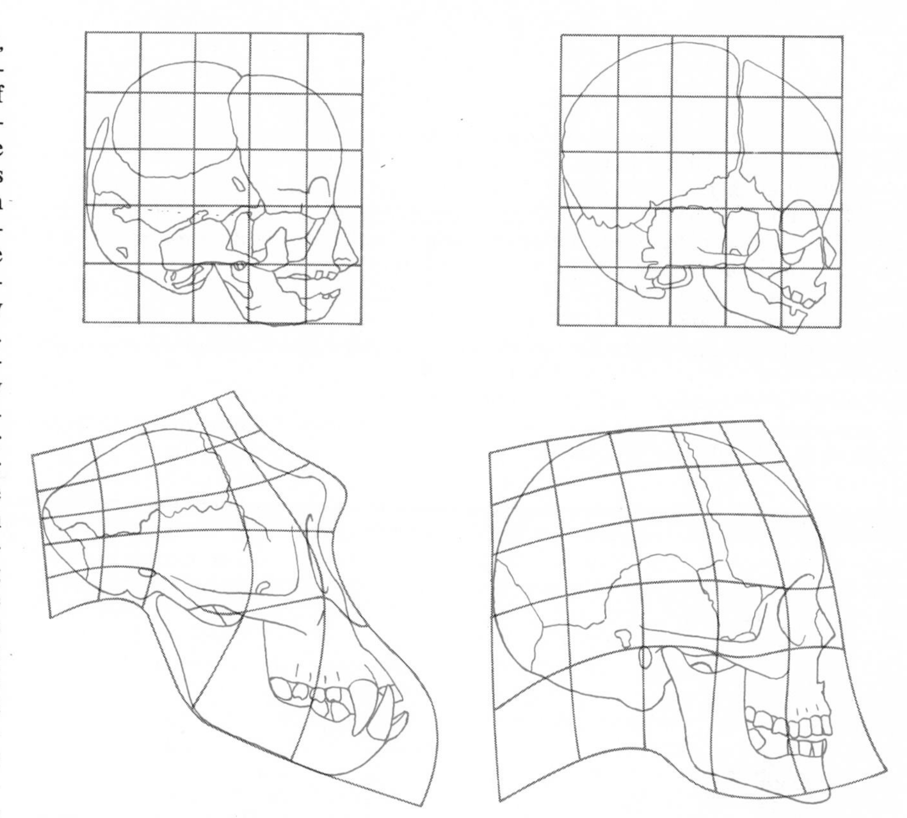

NEOTENY OF HUMAN SKULL is evident when the growth of the chimpanzee skull (*left*) and of the human skull (*right*) is plotted on transformed coordinates, which show the relative displacement of each part. The chimpanzee and the human skulls are much more similar at the fetal stage (*top*) than they are at the adult stage (*bottom*). The adult human skull also departs less from the fetal form than the adult chimpanzee skull departs from its fetal form, except in the case of the chin, which becomes relatively larger in human beings. The chin is a mental construct, however: the result of allometry, or differential growth, of different parts of human jaw.

Just such a compromise has been suggested for understanding the feeding behavior of some birds. Gordon H. Orians of the University of Washington studied the feeding behavior of birds that fly out from a nest, gather food and bring it back to the nest for consumption ("central-place foraging"). If the bird were to take food items indiscriminately as it came on them, the energy cost of the round trip from the nest and back might be greater than the energy gained from the food. On the other hand, if the bird chose only the largest food items, it might have to search so long that again the energy it consumed would be too great. For any actual distribution of food-particle sizes in nature there is some optimal foraging behavior for the bird that will maximize its net energy gain from feeding. Orians found that birds indeed do not take food particles at random but are biased in the direction of an optimal particle size. They do not, however, choose the optimal solution either. Orians' explanation was that the foraging behavior is a compromise between maximum energy efficiency and not staying away from the nest too long, because the young are exposed to predation when they are unattended.

The example of central-place foraging illustrates a basic assumption of all such engineering analyses, that of ceteris paribus, or all other things being equal. In order to make an argument that a trait is an optimal solution to a particular problem, it must be possible to view the trait and the problem in isolation, all other things being equal. If all other things are not equal, if a change in a trait as a solution to one problem changes the organism's relation to other problems of the environment, it becomes impossible to carry out the analysis part by part, and we are left in the hopeless position of seeing the whole organism as being adapted to the whole environment.

The mechanism by which organisms are said to adapt to the environment is that of natural selection. The theory of evolution by natural selection rests on three necessary principles: Different individuals within a species differ from one another in physiology, morphology and behavior (the principle of variation); the variation is in some way heritable, so that on the average offspring resemble their parents more than they resemble other individuals (the principle of heredity); different variants leave different numbers of offspring either immediately or in remote generations (the principle of natural selection).

These three principles are necessary and sufficient to account for evolutionary change by natural selection. There must be variation to select from; that variation must be heritable, or else there will be no progressive change from generation to generation, since there would be a random distribution of offspring even if some types leave more offspring than others. The three principles say nothing, however, about adaptation. In themselves they simply predict change caused by differential reproductive success without making any prediction about the fit of organisms to an ecological niche or the solution of ecological problems.

Adaptation was introduced by Darwin into evolutionary theory by a fourth principle: Variations that favor an individual's survival in competition with other organisms and in the face of environmental stress tend to increase reproductive success and so tend to be preserved (the principle of the struggle for existence). Darwin made it clear that the struggle for existence, which he derived from Thomas Malthus' *An Essay on the Principle of Population,* included more than the actual competition of two organisms for the same resource in short supply. He wrote: "I should premise that I use the term Struggle for Existence in a large and metaphorical sense.... Two canine animals in a time of dearth, may be truly said to struggle with each other which shall get food and live. But a plant on the edge of the desert is said to struggle for life against the drought."

The diversity that is generated by various mechanisms of reproduction and mutation is in principle random, but the diversity that is observed in the real world is nodal: organisms have a finite number of morphologies, physiologies and behaviors and occupy a finite number of niches. It is natural selection, operating under the pressures of the struggle for existence, that creates the nodes. The nodes are "adaptive peaks," and the species or other form occupying a peak is said to be adapted.

More specifically, the struggle for existence provides a device for predicting which of two organisms will leave more offspring. An engineering analysis can determine which of two forms of zebra can run faster and so can more easily escape predators; that form will leave more offspring. An analysis might predict the eventual evolution of zebra locomotion even in the absence of existing differences among individuals, since a careful engineer might think of small improvements in design that would give a zebra greater speed.

When adaptation is considered to be the result of natural selection under the pressure of the struggle for existence, it is seen to be a relative condition rather than an absolute one. Even though a species may be surviving and numerous, and therefore may be adapted in an absolute sense, a new form may arise that has a greater reproductive rate on the same resources, and it may cause the extinction of the older form. The concept of relative adaptation removes the apparent tautology in the theory of natural selection. Without it the theory of natural selection states that fitter individuals have more offspring and then defines the fitter as being those that leave more offspring; since some individuals will always have more offspring than others by sheer chance, nothing is explained. An analysis in which problems of design are posed and characters are understood as being design solutions breaks through this tautology by predicting in advance which individuals will be fitter.

The relation between adaptation and natural selection does not go both ways. Whereas greater relative adaptation leads to natural selection, natural selection does not necessarily lead to greater adaptation. Let us contrast two evolutionary scenarios. We begin with a resource-limited population of 100 insects of type *A* requiring one unit of food resource per individual. A mutation to a new type *a* arises that doubles the fecundity of its bearers but does absolutely nothing to the efficiency of the utilization of resources. We can calculate what happens to the composition, size and growth rate of the population over a period of time [*see illustration below*]. In a second scenario we again begin with the population of 100 individuals of type *A*, but now there arises a different mutation *a*, which does nothing to the fecundity of its bearers but doubles their efficiency of resource utilization. Again we can calculate the population history.

In both cases the new type *a* replaces the old type *A*. In the case of the first mutation nothing changes but the fecundity; the adult population size and the growth rate are the same throughout the process and the only effect is that twice as many immature stages are being produced to die before adulthood. In the second case, on the other hand, the population eventually doubles its adult members as well as its immature members, but not its fecundity. In the course of its evolution the second population has a growth rate greater than 1 for a while but eventually attains a constant size and stops growing.

In which of these populations, if in either, would the individuals be better

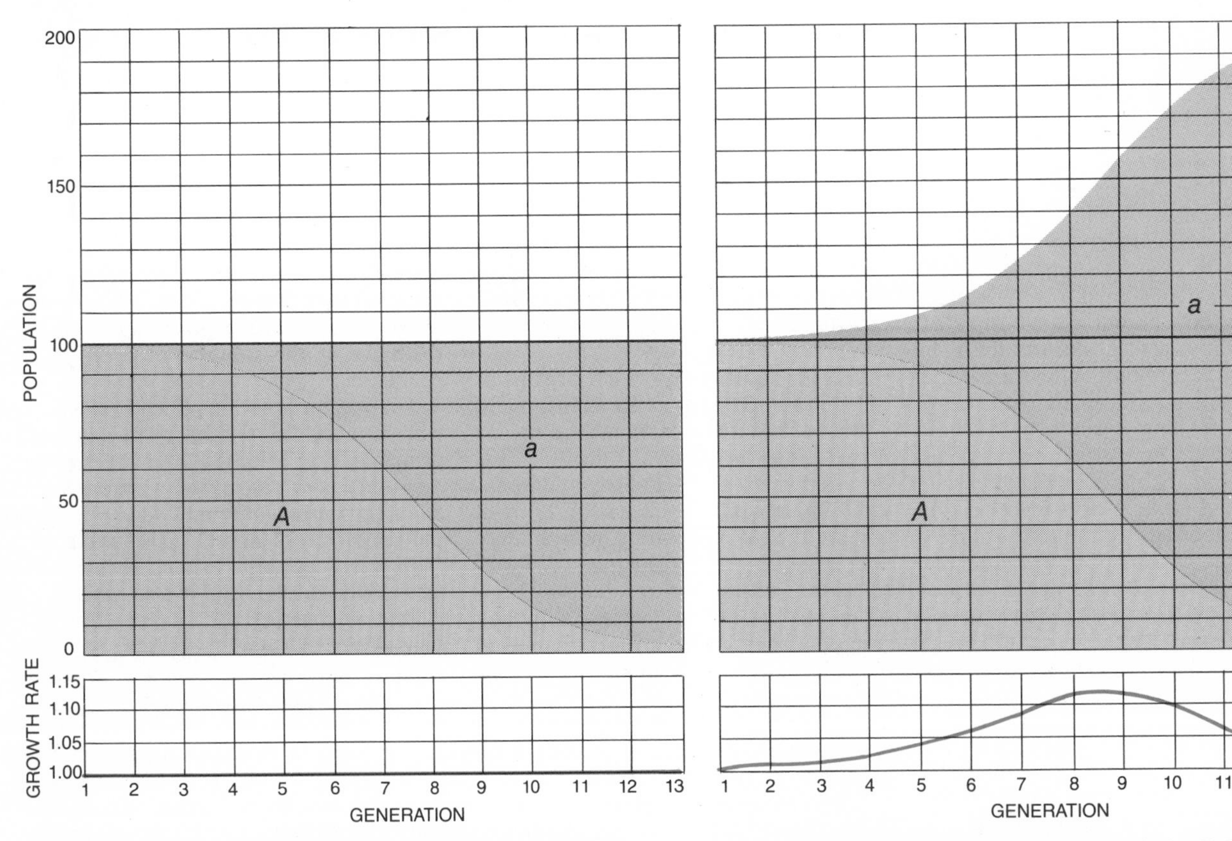

TWO DIFFERENT MUTATIONS have different demographic results for a resource-limited population of 100 insects. In one case (*left*) a mutation arises that doubles the fecundity of its bearers. The new type (*a*) replaces the old type (*A*), but the total population does not increase: the growth rate (*bottom*) remains 1.00. In the other case (*right*) a mutation arises that doubles the carrier's efficiency of resource utilization. Now the new population grows more rapidly, but only for a short time: eventually the growth rate falls back to 1.00 and the total population is stabilized at 200. The question is: Has either mutation given rise to a population that is better adapted?

adapted than those in the old population? Those with higher fecundity would be better buffered against accidents such as sudden changes in temperature since there would be a greater chance that some of their eggs would survive. On the other hand, their offspring would be more susceptible to the epidemic diseases of immature forms and to predators that concentrate on the more numerous immature forms. Individuals in the second population would be better adapted to temporary resource shortages, but also more susceptible to predators or epidemics that attack adults in a density-dependent manner. Hence there is no way we can predict whether a change due to natural selection will increase or decrease the adaptation in general. Nor can we argue that the population as a whole is better off in one case than in another. Neither population continues to grow or is necessarily less subject to extinction, since the larger number of immature or adult stages presents the same risks for the population as a whole as it does for individual families.

Unfortunately the concept of relative adaptation also requires the ceteris paribus assumption, so that in practice it is not easy to predict which of two forms will leave more offspring. A zebra having longer leg bones that enable it to run faster than other zebras will leave more offspring only if escape from predators is really the problem to be solved, if a slightly greater speed will really decrease the chance of being taken and if longer leg bones do not interfere with some other limiting physiological process. Lions may prey chiefly on old or injured zebras likely in any case to die soon, and it is not even clear that it is speed that limits the ability of lions to catch zebras. Greater speed may cost the zebra something in feeding efficiency, and if food rather than predation is limiting, a net selective disadvantage might result from solving the wrong problem. Finally, a longer bone might break more easily, or require greater developmental resources and metabolic energy to produce and maintain, or change the efficiency of the contraction of the attached muscles. In practice relative-adaptation analysis is a tricky game unless a great deal is known about the total life history of an organism.

Not all evolutionary change can be understood in terms of adaptation. First, some changes will occur directly by natural selection that are not adaptive, as for example the changes in fecundity and feeding efficiency in the hypothetical example I cited above.

Second, many changes occur indirectly as the result of allometry, or differential growth. The rates of growth of different parts of an organism are different,

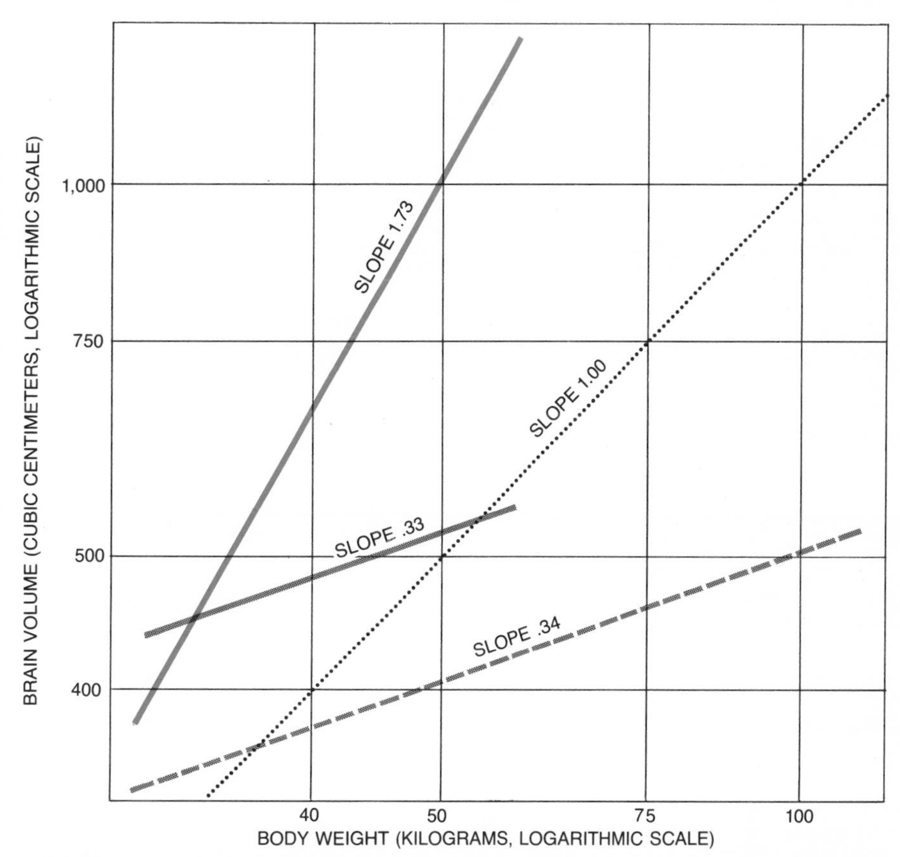

ALLOMETRY, or differential growth rates for different parts, is responsible for many evolutionary changes. Allometry is illustrated by this comparison of the ratio of brain size to body weight in a number of species of the pongids, or great apes (*broken black curve*), of *Australopithecus*, an extinct hominid line (*solid black*), and of hominids leading to modern man (*color*). A slope of less than 1.00 means the brain has grown more slowly than the body. The slope of more than 1.00 for the human lineage indicates a clear change in the evolution of brain size.

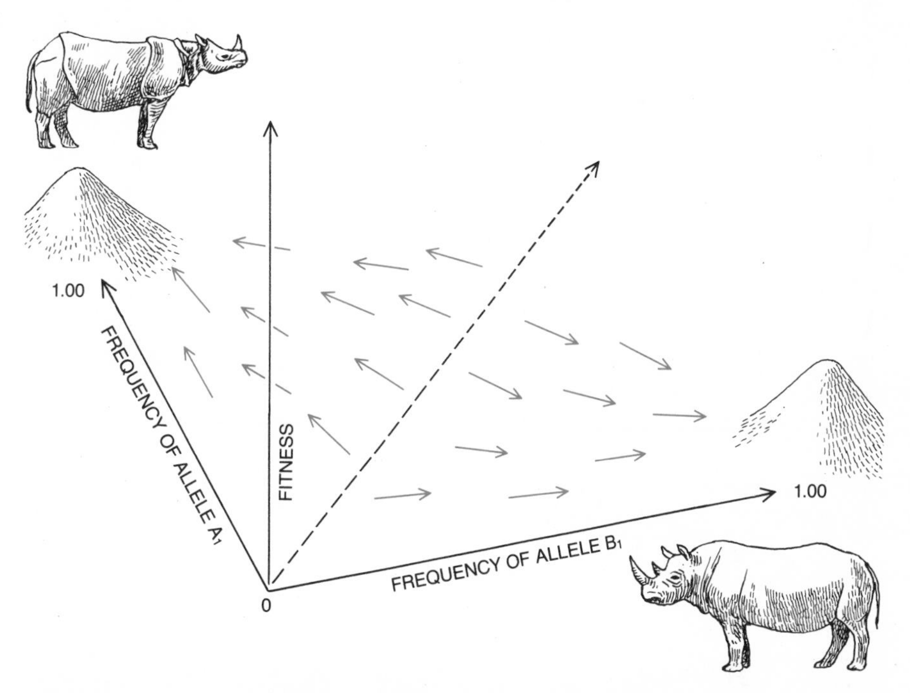

ALTERNATIVE EVOLUTIONARY PATHS may be taken by two species under similar selection pressures. The Indian rhinoceros has one horn and the African rhinoceros has two horns. The horns are adaptations for protection in both cases, but the number of horns does not necessarily constitute a specifically adaptive difference. There are simply two adaptive peaks in a field of gene frequencies, or two solutions to the same problem; some variation in the initial conditions led two rhinoceros populations to respond to similar pressures in different ways. For each of two hypothetical genes there are two alleles: A_1 and A_2, B_1 and B_2. A population of genotype A_1B_2 has one horn and a population of genotype A_2B_1 has two horns.

so that large organisms do not have all their parts in the same proportion. This allometry shows up both between individuals of the same species and between species. Among primate species the brain increases in size more slowly than the body; small apes have a proportionately larger brain than large apes. Since the differential growth is constant for all apes, it is useless to seek an adaptive reason for gorillas' having a relatively smaller brain than, say, chimpanzees.

Third, there is the phenomenon of pleiotropy. Changes in a gene have many different effects on the physiology and development of an organism. Natural selection may operate to increase the frequency of the gene because of one of the effects, with pleiotropic, or unrelated, effects being simply carried along. For example, an enzyme that helps to detoxify poisonous substances by converting them into an insoluble pigment will be selected for its detoxification properties. As a result the color of the organism will change, but no adaptive explanation of the color per se is either required or correct.

Fourth, many evolutionary changes may be adaptive and yet the resulting differences among species in the character may not be adaptive; they may simply be alternative solutions to the same problem. The theory of population genetics predicts that if more than one gene influences a character, there may often be several alternative stable equilibriums of genetic composition even when the force of natural selection remains the same. Which of these adaptive peaks in the space of genetic composition is eventually reached by a population depends entirely on chance events at the beginning of the selective process. (An exact analogy is a pinball game. Which hole the ball will fall into under the fixed force of gravitation depends on small variations in the initial conditions as the ball enters the game.) For example, the Indian rhinoceros has one horn and the African rhinoceros has two. Horns are an adaptation for protection against predators, but it is not true that one horn is specifically adaptive under Indian conditions as opposed to two horns on the African plains. Beginning with two somewhat different developmental systems, the two species responded to the same selective forces in slightly different ways.

Finally, many changes in evolution are likely to be purely random. At the present time population geneticists are sharply divided over how much of the evolution of enzymes and other molecules has been in response to natural selection and how much has resulted from the chance accumulation of mutations. It has proved remarkably difficult to get compelling evidence for changes in enzymes brought about by selection, not to speak of evidence for adaptive changes; the weight of evidence at present is that a good deal of amino acid substitution in evolution has been the result of the random fixation of mutations in small populations. Such random fixations may in fact be accelerated by natural selection if the unselected gene is genetically linked with a gene that is undergoing selection. The unselected gene will then be carried to high frequency in the population as a "hitchhiker."

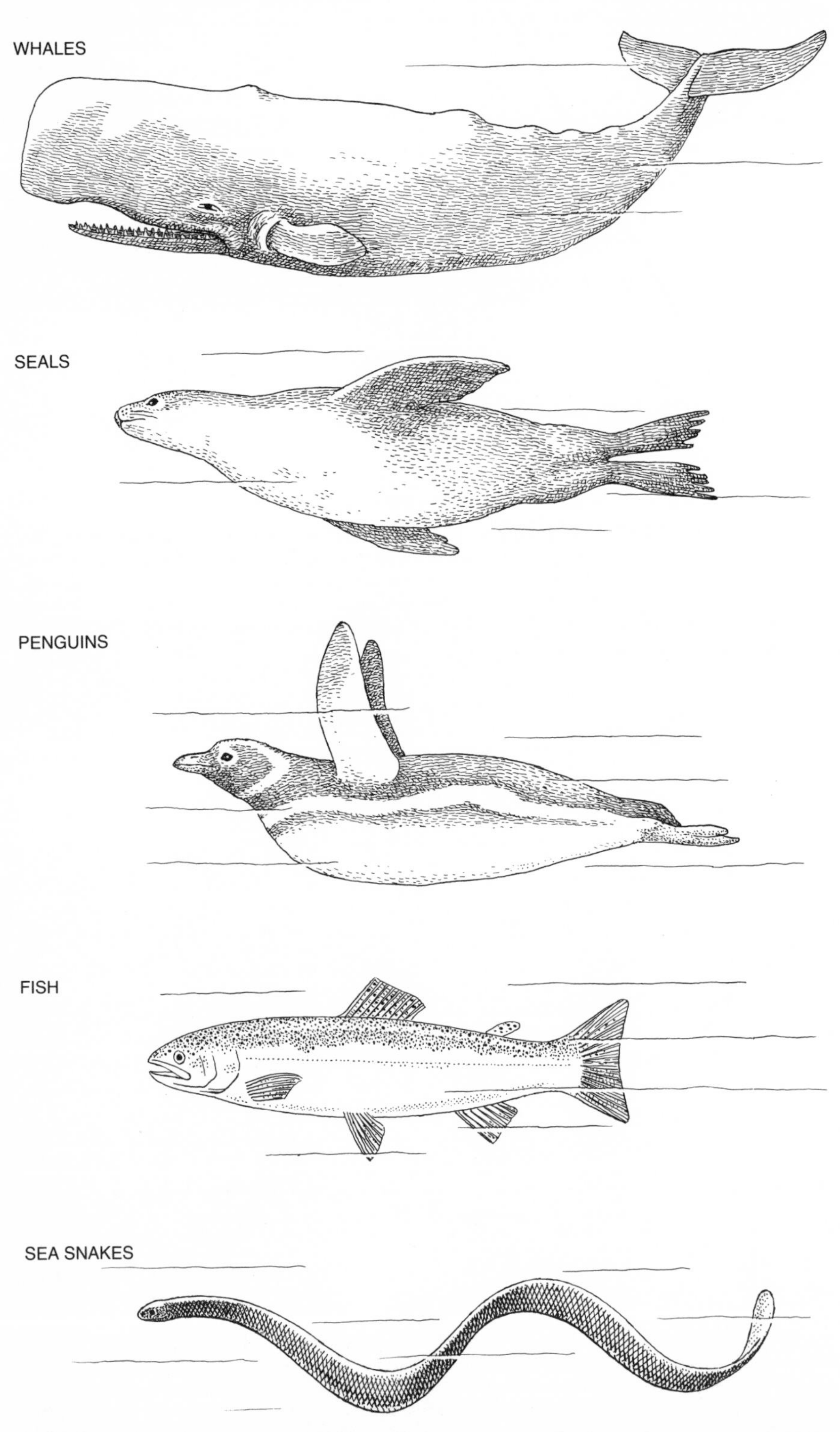

REALITY OF ADAPTATION is demonstrated by the indisputable fact that unrelated groups of animals do respond to similar selective pressures with similar adaptations. Locomotion in water calls for a particular kind of structure. And the fact is that whales and seals have flippers and flukes, penguins have paddles, fish have fins and sea snakes have a flat cross section.

If the adaptationist program is so fraught with difficulties and if there are so many alternative explanations of evolutionary change, why do biologists not abandon the program altogether?

There are two compelling reasons. On the one hand, even if the assertion of universal adaptation is difficult to test because simplifying assumptions and ingenious explanations can almost always result in an ad hoc adaptive explanation, at least in principle some of the assumptions can be tested in some cases. A weaker form of evolutionary explanation that explained some proportion of the cases by adaptation and left the rest to allometry, pleiotropy, random gene fixations, linkage and indirect selection would be utterly impervious to test. It would leave the biologist free to pursue the adaptationist program in the easy cases and leave the difficult ones on the scrap heap of chance. In a sense, then, biologists are forced to the extreme adaptationist program because the alternatives, although they are undoubtedly operative in many cases, are untestable in particular cases.

On the other hand, to abandon the notion of adaptation entirely, to simply observe historical change and describe its mechanisms wholly in terms of the different reproductive success of different types, with no functional explanation, would be to throw out the baby with the bathwater. Adaptation is a real phenomenon. It is no accident that fish have fins, that seals and whales have flippers and flukes, that penguins have paddles and that even sea snakes have become laterally flattened. The problem of locomotion in an aquatic environment is a real problem that has been solved by many totally unrelated evolutionary lines in much the same way. Therefore it must be feasible to make adaptive arguments about swimming appendages. And this in turn means that in nature the ceteris paribus assumption must be workable.

It can only be workable if both the selection between character states and reproductive fitness have two characteristics: continuity and quasi-independence. Continuity means that small changes in a characteristic must result in only small changes in ecological relations; a very slight change in fin shape cannot cause a dramatic change in sexual recognition or make the organism suddenly attractive to new predators. Quasi-independence means that there is a great variety of alternative paths by which a given characteristic may change, so that some of them will allow selection to act on the characteristic without altering other characteristics of the organism in a countervailing fashion; pleiotropic and allometric relations must be changeable. Continuity and quasi-independence are the most fundamental characteristics of the evolutionary process. Without them organisms as we know them could not exist because adaptive evolution would have been impossible.

The Authors
Bibliography
Index

The Authors

ERNST MAYR ("Evolution") is professor emeritus of zoology at Harvard University. Born in Germany, he was educated at the University of Berlin, where he received his Ph.D. in zoology in 1926. After six years as assistant curator of the zoological museum there, he came to the U.S. to become associate curator of the Whitney-Rothschild Collection of the American Museum of Natural History in New York. He joined the Harvard faculty in 1953 and served as director of the Museum of Comparative Zoology from 1961 through 1970; he has been professor emeritus since 1975. In addition to evolutionary theory Mayr's research has covered ornithology, systematics and the history and philosophy of biology. He is the author of a number of books, including *Principles of Systematic Zoology* (1969) and *Populations, Species, and Evolution* (1970).

FRANCISCO J. AYALA ("The Mechanisms of Evolution") is professor of genetics at the University of California at Davis and director of the Institute of Ecology there. Born and educated in Madrid, he came to the U.S. in 1961 to study genetics and evolution with Theodosius Dobzhansky at Columbia University, obtaining his Ph.D. in 1964. After an appointment at Rockefeller University he moved to Davis in 1971. His scientific career has centered on the application of molecular biology to the study of evolution, particularly with regard to measurements of genetic variation in natural populations, of rates of evolution and of the amount of genetic change involved in the formation of new species. His other professional interests include population ecology and philosophical and ethical questions related to biology. Ayala is the author of *Molecular Evolution* (1976) and coauthor of *Evolution* (1977), a statement of the current status of evolutionary theory.

RICHARD E. DICKERSON ("Chemical Evolution and the Origin of Life") is professor of chemistry at the California Institute of Technology. He received his bachelor's degree at the Carnegie Institute of Technology and his Ph.D. in physical chemistry from the University of Minnesota in 1957. After postdoctoral training at the University of Cambridge, where he worked with J. C. Kendrew on the first high-resolution X-ray analysis of the structure of a crystalline protein (sperm-whale myoglobin), he spent four years on the faculty of the University of Illinois, moving to Cal Tech in 1963. He and his co-workers have solved the structures of the digestive enzyme trypsin and cytochrome *c* from the horse, the tuna and two different bacteria; they are now working on the structure of the repressor-operator complex for the *lac* operon. Dickerson is the author or coauthor of six textbooks in chemistry, biochemistry and biology.

J. WILLIAM SCHOPF ("The Evolution of the Earliest Cells") is professor of paleobiology at the University of California at Los Angeles. He did his undergraduate work at Oberlin College and obtained his Ph.D. in biology from Harvard University in 1968, thereafter joining the U.C.L.A. faculty. Schopf's research on the earliest fossil records of life has involved fieldwork in North and South America, Australia, India, the U.S.S.R. and the People's Republic of China.

JAMES W. VALENTINE ("The Evolution of Multicellular Plants and Animals") is professor of geological sciences at the University of California at Santa Barbara. He has spent most of his professional life in the University of California system, both as a graduate student on the Los Angeles campus (where he received his Ph.D. in geology in 1958) and as a teacher on the Davis and Santa Barbara campuses. His early research was on the association and distribution patterns of fossil marine mollusks, but with the rise of the theory of plate tectonics he became interested in the general relations between evolution and the dynamic history of the earth. Valentine writes: "My chief hobby is collecting the writings of Charles Darwin, all issues in all languages, but most of my spare time is now spent with my wife in remodeling our new home in the Santa Ynez valley, converted from a horse barn."

ROBERT M. MAY ("The Evolution of Ecological Systems") is professor of zoology at Princeton University and chairman of the University Research Board. He was born and educated in Australia and obtained his Ph.D. in theoretical physics from the University of Sydney in 1960. After two years as Gordon Mackay Lecturer in the division of engineering and applied mathematics at Harvard University he returned to Sydney as senior lecturer (and later professor) in theoretical physics. In 1971–72, during a sabbatical year spent at the University of Oxford and at the Institute for Advanced Study in Princeton, a series of events turned his interests to the organization of plant and animal communities, and he moved to Princeton as professor of biology in 1973. May's current research interests include the population dynamics of insects, harvested fishes and whales, and the overall transmission cycle of tropical diseases such as schistosomiasis. With his wife Judith, May enjoys "walking in Britain and running in America."

JOHN MAYNARD SMITH ("The Evolution of Behavior") is professor of biology at the University of Sussex. He initially studied engineering at the University of Cambridge and worked during World War II as an aircraft engineer. After the war he did graduate work in zoology at University College London under J. B. S. Haldane and received his Ph.D. in 1950. For the next 15 years he was lecturer in zoology at University College, moving in 1965 to the new University of Sussex, where he founded the School of Biological Sciences. Smith's present interests include the evolution of behavior and of sexual reproduction.

SHERWOOD L. WASHBURN ("The Evolution of Man") is professor of physical anthropology at the University of California at Berkeley. He did his undergraduate and graduate work at Harvard University, obtaining his Ph.D. in 1940. He then taught human anatomy at the Columbia University College of Physicians and Surgeons, moving in 1947 to the University of Chicago. It was there he began to supplement his laboratory investigations of human evolution with field studies of primates in their natural habitat, a new approach that soon acquired a considerable following. He joined the Berkeley faculty in 1958. Washburn has written several monographs and books on human evolution, of which the most recent is *Ape into Man: A Study of Human Evolution* (1974) with Ruth Moore.

RICHARD C. LEWONTIN ("Adaptation") is Alexander Agassiz Professor of Zoology at Harvard University and professor of population sciences at the Harvard School of Public Health. He was educated at Harvard College and Columbia University, where he received his Ph.D. in zoology in 1954. After sojourns at North Carolina State College, the University of Rochester, the University of Chicago (where he was associate dean of the division of biological sciences) and Syracuse University, he joined the Harvard faculty in 1973. He was elected to the National Academy of Sciences in 1968 but resigned three years later "on a question of political principle." He writes: "At present I am working on a problem that has preoccupied me for the past 25 years: what the nature of genetic variation in natural populations is and how it is controlled." Lewontin's other research includes theoretical and computer-simulation studies of the epidemiology of the tropical disease schistosomiasis in the hope of devising effective control measures.

Bibliography

Readers interested in further explanation of the subjects covered by the articles in this book may find the following lists of publications helpful.

EVOLUTION

ON THE ORIGIN OF SPECIES. Charles Darwin. Facsimile Edition. Harvard University Press, 1964.

THE NATURE OF THE DARWINIAN REVOLUTION. Ernst Mayr in *Science,* Vol. 176, No. 4038, pages 981–989; June 2, 1972.

DARWIN AND NATURAL SELECTION. Ernst Mayr in *American Scientist,* Vol. 65, No. 3, pages 321–327; May–June, 1977.

THE MECHANISMS OF EVOLUTION

CONSTRUCTION OF PHYLOGENETIC TREES. Walter M. Fitch and Emanuel Margoliash in *Science,* Vol. 155, No. 3760, pages 279–284; January 20, 1967.

THE GENETIC BASIS OF EVOLUTIONARY CHANGE. Richard C. Lewontin. Columbia University Press, 1974.

DARWINIAN EVOLUTION IN THE GENEALOGY OF HAEMOGLOBIN. Morris Goodman, G. William Moore and Genji Matsuda in *Nature,* Vol. 253, No. 5493, pages 603–608; February 20, 1975.

MOLECULAR EVOLUTION. Edited by Francisco J. Ayala. Sinauer Associates, Inc., 1976.

EVOLUTION. Theodosius Dobzhansky, Francisco J. Ayala, G. Ledyard Stebbins and James W. Valentine. W. H. Freeman and Company, 1977.

CHEMICAL EVOLUTION AND THE ORIGIN OF LIFE

THE CHEMICAL ELEMENTS IN NATURE. Frank Henry Day. G. G. Harrap, 1963.

THE CHEMICAL ELEMENTS OF LIFE. Earl Frieden in *Scientific American,* Vol. 227, No. 1, pages 52–60; July, 1972.

CHEMISTRY, MATTER, AND THE UNIVERSE. Richard E. Dickerson and Irving Geis. W. A. Benjamin, Inc., 1976.

THE EVOLUTION OF THE EARLIEST CELLS

ORIGIN OF EUKARYOTIC CELLS. Lynn Margulis. Yale University Press, 1970.

THE BIOLOGY OF BLUE-GREEN ALGAE. Edited by N. G. Carr and B. A. Whitton. University of California Press, 1973.

PRECAMBRIAN PALEOBIOLOGY: PROBLEMS AND PERSPECTIVES. J. William Schopf in *Annual Review of Earth and Planetary Sciences: Vol. 3,* edited by Fred A. Donath, Francis G. Stehli and George W. Wetherill. Annual Reviews Inc., 1975.

BIOSTRATIGRAPHIC USEFULNESS OF STROMATOLITIC PRECAMBRIAN MICROBIOTAS: A PRELIMINARY ANALYSIS. J. William Schopf in *Precambrian Research,* Vol. 5, No. 2, pages 143–173; August, 1977.

THE EVOLUTION OF MULTICELLULAR PLANTS AND ANIMALS

EVOLUTIONARY PALEOECOLOGY OF THE MARINE BIOSPHERE. James W. Valentine. Prentice-Hall, Inc., 1973.

PLATE TECTONICS AND THE HISTORY OF LIFE IN THE OCEANS. James W. Valentine and Eldridge M. Moores in *Scientific American,* Vol. 230, No. 4, pages 80–89; April, 1974.

THE EVOLVING CONTINENTS. Brian F. Windley. John Wiley & Sons, Inc., 1977.

PRINCIPLES OF PALEONTOLOGY. David M. Raup and Steven M. Stanley. W. H. Freeman and Company, 1978.

THE EVOLUTION OF ECOLOGICAL SYSTEMS

GEOGRAPHICAL ECOLOGY: PATTERNS IN THE DISTRIBUTION OF SPECIES. Robert H. MacArthur. Harper and Row, 1972.

STABILITY AND COMPLEXITY IN MODEL ECOSYSTEMS. Robert M. May. Princeton University Press, 1973.

COMMUNITIES AND ECOSYSTEMS. Robert H. Whittaker. The Macmillan Company, 1975.

ECOLOGY AND EVOLUTION OF COMMUNITIES. Edited by Martin L. Cody and Jared M. Diamond. Harvard University Press, 1975.

THEORETICAL ECOLOGY: PRINCIPLES AND APPLICATIONS. Edited by Robert M. May. W. B. Saunders Company, 1976.

THE EVOLUTION OF BEHAVIOR

THE GENETICAL EVOLUTION OF SOCIAL BEHAVIOUR. W. D. Hamilton in *Journal of Theoretical Biology,* Vol. 7, pages 1–52; 1964.

THE SELFISH GENE. Richard Dawkins. Oxford University Press, 1976.

THE EVOLUTION OF MAN

PHYLOGENY OF THE PRIMATES: A MULTIDISCIPLINARY APPROACH. Edited by W. Patrick Luckett and Frederick S. Szalay. Plenum Press, 1975.

EVOLUTION AT TWO LEVELS IN HUMANS AND CHIMPANZEES. Mary-Claire King and A. C. Wilson in *Science,* Vol. 188, No. 4184, pages 107–116; April 11, 1975.

Human Origins: Louis Leakey and the East African Evidence. Edited by Glynn Ll. Isaac and Elizabeth R. McCown. W. A. Benjamin, Inc., 1976.

Molecular Anthropology: Evolving Information Molecules in the Ascent of the Primates. Edited by Morris Goodman and Richard E. Tashian. Plenum Press, 1976.

Origins. Richard Leakey and Roger Lewin. E. P. Dutton & Co., 1977.

Human Evolution: Biosocial Perspectives. Edited by Sherwood L. Washburn and Elizabeth R. McCown. Benjamin/Cummings Publishing Company, 1978.

ADAPTATION

Adaptation and Natural Selection: A Critique of Some Current Evolutionary Thought. George C. Williams. Princeton University Press, 1966.

Evolution in Changing Environments: Some Theoretical Explorations. Richard Levins. Princeton University Press, 1968.

Adaptation and Diversity: Natural History and the Mathematics of Evolution. E. G. Leigh, Jr. Freeman, Cooper and Company, 1971.

Index

5051 002